An Introduction to Stellar Magnetic Activity

AAS Editor in Chief

Ethan Vishniac, Johns Hopkins University, Maryland, USA

About the program:

AAS-IOP Astronomy ebooks is the official book program of the American Astronomical Society (AAS) and aims to share in depth the most fascinating areas of astronomy, astrophysics, solar physics, and planetary science. The program includes publications in the following topics:

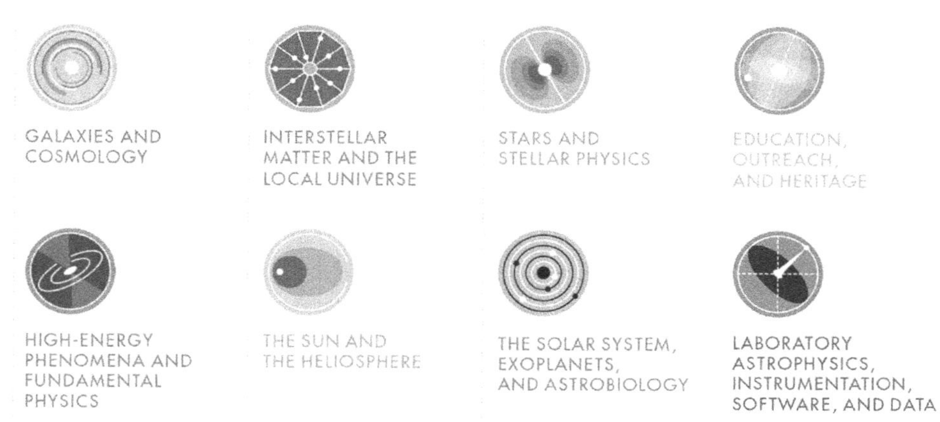

Books in the program range in level from short introductory texts on fast-moving areas, graduate and upper-level undergraduate textbooks, research monographs, and practical handbooks.

For a complete list of published and forthcoming titles, please visit iopscience.org/books/aas.

About the American Astronomical Society

The American Astronomical Society (aas.org), established 1899, is the major organization of professional astronomers in North America. The membership (~7,000) also includes physicists, mathematicians, geologists, engineers, and others whose research interests lie within the broad spectrum of subjects now comprising the contemporary astronomical sciences. The mission of the Society is to enhance and share humanity's scientific understanding of the universe.

Editorial Advisory Board

An Introduction to Stellar Magnetic Activity

Gibor Basri
University of California, Berkeley, USA

IOP Publishing, Bristol, UK

ISBN   978-0-7503-2132-7 (ebook)
ISBN   978-0-7503-2130-3 (print)
ISBN   978-0-7503-2133-4 (myPrint)
ISBN   978-0-7503-2131-0 (mobi)

DOI   10.1088/2514-3433/ac2956

Version: 20220101

AAS–IOP Astronomy
ISSN 2514-3433 (online)
ISSN 2515-141X (print)

British Library Cataloguing-in-Publication Data: A catalogue record for this book is available from the British Library.

Published by IOP Publishing, wholly owned by The Institute of Physics, London

IOP Publishing, Temple Circus, Temple Way, Bristol, BS1 6HG, UK

US Office: IOP Publishing, Inc., 190 North Independence Mall West, Suite 601, Philadelphia, PA 19106, USA

Cover image: Sun Emits a Solstice CME. Image credit: NASA Goddard/GSFC.

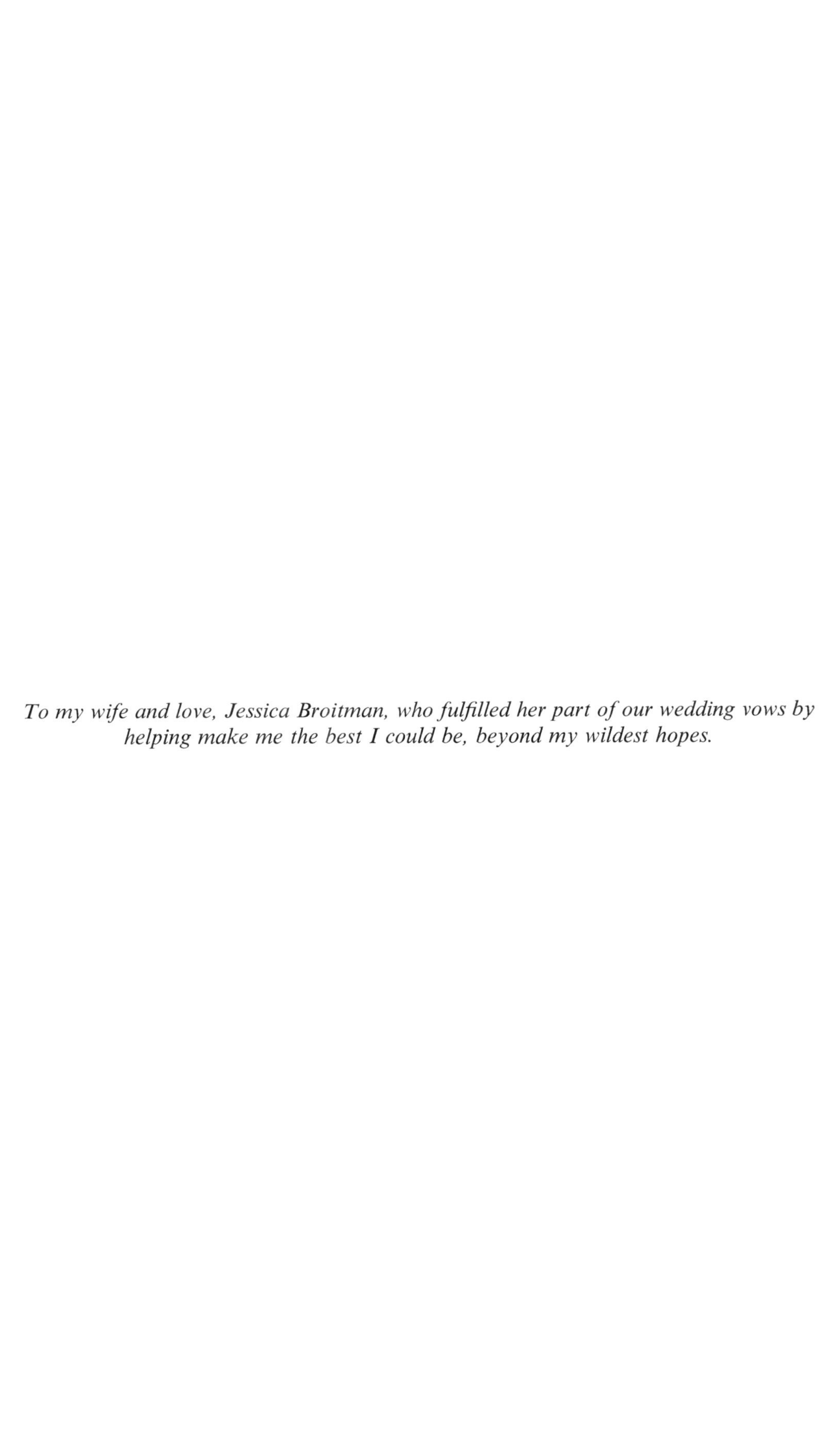

To my wife and love, Jessica Broitman, who fulfilled her part of our wedding vows by helping make me the best I could be, beyond my wildest hopes.

Contents

Preface

This book is written as an introduction to the field of stellar magnetic activity for advanced undergraduate or beginning graduate students. The subject began with the only star whose surface and atmosphere we can see in detail—our Sun—so it is a good touchstone to start from. The primary focus of the book is that type of magnetic activity, but in the context of stars of all masses and ages that produce magnetic fields in their interiors. Topics include the production and measurements of the fields themselves, their effects on the stellar atmosphere and the diagnostics we use to understand them, and the effects of the magnetic activity on the star and its surrounds over time.

I have tried to summarize the state of the field in 2020 with as much brevity as I could manage. The book is partly intended as a history of how we arrived where we are today, so it contains early pioneering references as well as the latest work at the time of writing. Sometimes the narrative is chronological but often it proceeds through a physical description of phenomena on a path that made sense to me. I particularly hope to impart the basic physical intuition needed to understand the ideas and intent behind investigations of stellar magnetic activity, to provide an overview of the various observations, theories, and methods of analysis that have been employed, and to summarize the basic results and relations they have produced.

I assume that the reader is familiar with electromagnetism and quantum mechanics at the level of college physics, and has had the equivalent of an introductory astronomy course. The appendix on radiative transfer has the most equations and is more technical. The reader can skip it if they already know what optical depth and source functions are, although I recommend at least skimming it. Those that want to understand how we infer physical information from the strong spectral lines diagnostic of stellar activity will benefit from a more careful reading of it.

I have deliberately not written either an extensive review of the literature or a textbook that contains the mathematics and extensive physical details behind much of the science, so there are few equations. The book does not contain a comprehensive bibliography, so the references cited are chosen in part because they provide useful further references or very illustrative figures. I have also tried to point to more extensive reviews on various topics. I apologize in advance to my many friends and colleagues and other authors whose papers weren't explicitly cited despite their excellent relevance.

Hopefully this book provides an overview of the whole field with enough observational and physical context for the knowledge we have gained that the reader learns where and for what purpose to delve into the literature for more details. There will continue to be rapid advances, so this is supposed to be an initial resource for those new to the area or an overview for those already involved who want a wider or historical perspective. I have indicated directions that new research might be headed when I could. Ideally the book might help stimulate the interest of new scientists, and the launch of new investigations that push the field forward.

Gibor Basri
Berkeley, CA
August 2021

Acknowledgements

It would be impossible for me to list here all the folks who have mentored, collaborated, or been friends or colleagues with me on my journey through a career on stellar activity. I owe the largest debt to my thesis advisor, Jeff Linsky. He not only got me started in astrophysics but also, to promote my career, refused to let me stay as a postdoc in Boulder. That was the impetus that led me to Berkeley and my academic home. At Berkeley I particularly acknowledge Stu Bowyer who first invited me, Len Kuhi who got me into star formation, and Frank Shu who solidified my interest in that. The Berkeley Astronomy Department proved a congenial and productive base for my entire faculty career and I thank all my faculty colleagues for making it so. My graduate students, postdocs, and primary collaborators are the central cause of the great satisfaction I have experienced in my professional life.

Author Biography

Gibor Basri

Gibor Basri earned a BSc in Physics from Stanford University in 1973, and a PhD in Astrophysics from the Univ. of Colorado, Boulder in 1979. His thesis was on ultraviolet observations and semi-empirical models of the chromospheres of the Sun and stars. An award of a Chancellor's Postdoctoral Fellowship then brought him to the Univ. of California, Berkeley. He joined the faculty of the UC Berkeley Astronomy Dept. in 1982.

His research in the 1980s concentrated on star formation and the study of T Tauri stars, as well as continuing studies of stellar activity. In the 1990s he continued work on these topics, including a number of discoveries about rotation and activity on low mass stars and development of methods for directly measuring magnetic fields on stars. In 1995 he was a discoverer of and became a world expert on brown dwarfs. That work included his invention of "lithium dating," which increased the inferred ages of young clusters like the Pleiades significantly. He wrote an *Annual Reviews of Astronomy and Astrophysics* article on "Observations of Brown Dwarfs" in 2000 and was lead author on an *Annual Reviews of Earth and Planetary Science* article in 2006 entitled "Brown Dwarfs to Planetesimals: What is a Planet?." In the 2000s he became a Co-Investigator on NASA's Kepler mission, which revolutionized our understanding of exoplanets and took stellar science in new directions. His research currently focuses on extracting information about starspots from Kepler light curves.

Prof. Basri is passionate about creating greater diversity within the physical sciences, including in outreach programs, research mentorship programs, and on policies about faculty hiring and retention. This and other relevant work eventually led him to be chosen as UC Berkeley's founding Vice Chancellor for Equity and Inclusion from 2007–2015. He is now Professor Emeritus of Astronomy.

An Introduction to Stellar Magnetic Activity

Gibor Basri

Chapter 1

Introduction to the Magnetic Sun

To the human eye (with safeguards), the Sun is usually a featureless disk of light. This visible disk is the part of the solar atmosphere called the photosphere. It is the "surface" of the Sun, the location from which most of its energy escapes into space as visible light. At certain times and places a few dark spots may also be seen on the solar disk (more commonly if using magnification). The spots themselves consist of a dark inner core (umbra) often surrounded by a less dark irregular ring (penumbra). Such sightings constitute the simplest observation of solar magnetic activity—these sunspots are due to very strong magnetic fields in confined regions on the Sun. With precise visible imagery it is also possible to see slightly brighter regions that are associated with somewhat strengthened magnetic fields. These "faculae" mark magnetic active regions, but are only seen toward the limb of the Sun because of geometrical effects related to convective granulation and opacity changes caused by the magnetic field. The photosphere is discussed in more detail in Chapter 2. As is the case with all topics in this book, we utilize the Sun as a starting point to provide context for what has been learned on other stars.

A much rarer opportunity to observe solar magnetic activity with the naked eye occurs during total solar eclipses, when a rim and/or loops of reddish color appear just above the solar limb. These parts of the solar atmosphere are actually hotter than the photosphere, contrary to what would be expected if energy were being transported outward purely by radiation once the atmosphere has become transparent. The red rim gets its name from its color (the "chromosphere"); the color is due to a transition of hydrogen that is excited in that part of the atmosphere. The loops ("prominences") are extensions of the chromosphere higher up along magnetic field lines rooted in the photosphere. The existence of the chromosphere is almost entirely due to the presence of magnetic fields (Chapter 3), although it is possible that there could be a weak "basal" chromosphere due to heating caused by the fact that acoustic energy from the convection below encounters a rapidly decreasing density structure and the acoustic waves could "break" creating sonic shock waves

doi:10.1088/2514-3433/ac2956ch1

that dissipate their energy. It is hard to observe the chromosphere with visible light images except at the eclipsed limb or in selected spectral lines; we return to it below when discussing other types of observations.

The chromosphere is terminated by a thin layer above it where the temperature jumps higher by two orders of magnitude. This is caused by the loss of neutral hydrogen (and other elements) that can provide radiative cooling, once the temperature rises much above 10,000 K. This thermal instability is not fixed until the temperature is above a million degrees, when new sources of cooling (like free–free emission from electrons) kick in. This layer is called the transition region. It cannot be seen in visible light; its emission is almost entirely in the ultraviolet (UV) and far ultraviolet (FUV) and extreme ultraviolet (EUV) parts of the spectrum. Above it, the pearly white corona provides a magnificent extended halo around the eclipsed Sun. It is the very hot and thin highly extended part of the solar atmosphere whose temperatures of millions of degrees are entirely due to the presence of magnetic fields, via some combination of magnetic waves and magnetic dissipation (the detailed heating mechanisms are not fully understood). These parts of of the Sun are discussed in Section 4.2.

The intrinsic coronal emission is at X-ray wavelengths that can only be observed from space; the visible light glow is just Thompson scattering of photospheric light off free electrons. The X-ray emission arises primarily from closed magnetic field loops rooted in the photosphere. The densities in these loops are highest for the ones rooted in active regions; the more field in the region the brighter the coronal loops tend to be. Because of turbulent motions in the photosphere, the loops are dynamic phenomena with ever-changing configurations. Sometimes this leads to massive explosions on the Sun due to sudden dissipation of magnetic fields, called "flares." The primary heating and emissions from these are extremely hot and high energy (and so must be observed in X-rays and γ-rays), but they also accelerate particles that stream down and heat the chromosphere and photosphere creating visible radiation. Flares actually produce strong emission across the entire electromagnetic spectrum. The disruption of coronal loops or prominences can also cause massive ejections of solar plasma into interplanetary space (called "coronal mass ejections" or CMEs). All these apparitions are direct manifestations of atmospheric structures and dynamics that would not be present at all without magnetic fields.

In some parts of the corona the field is open out to several solar radii and beyond. These regions are called "coronal holes" because their X-ray emission is much weaker; they look dark in contrast to the closed loop regions. The structure of the corona changes with the solar cycle, being relatively more symmetric (even mostly dipolar) during solar minimum and quite structured with helmet streamers and long linear structures closer to the equator during solar maximum. Open field regions are the primary source of the solar wind—the remaining part of the Sun that extends well beyond the planetary system until terminating at the interstellar medium. This flow of particles from the Sun is driven by the heat of the corona and so is another consequence of the solar magnetic field. In 1958 Eugene Parker realized that structure of the field and the decreasing gravity field would cause the wind to transition from subsonic to supersonic at a few solar radii, as if flowing through a

Laval nozzle, which was later observed to be the case. There are two general components to the solar wind, fast and slow, which attain velocities of roughly 700–800 and 300 km s^{-1} respectively near the Earth. The fast flows tend to originate in coronal holes. The fact the Sun is rotating also drags the wind structure into a trailing Archimedes spiral; at the Alfvén radius the field can no longer force the material to keep up with its photospheric roots. The solar wind impacts all the planets (or their magnetospheres if they have one) and is a major source of the aurorae on the Earth and other planets. These topics are expanded upon in Sections 4.4 and 7.5.

Of course, visible light is just a small part of the electromagnetic spectrum (albeit an important one to humans). It is not accidental that it is the part of the spectrum where most solar energy is emitted; eyes presumably evolved to take advantage of this. It should already be apparent that the visible is not the optimal part of the spectrum in which to study stellar magnetic activity. Because almost all magnetic features but sunspots are formed above the photosphere, they are basically transparent to visible light. This means that only the most opaque visible atomic spectral lines are suitable for observing the chromosphere in visible light images. Such images must be obtained with narrow-band filters to isolate the light in these lines.

By far the most commonly used lines are the Fraunhofer C, H, and K lines; strong lines in the visible solar spectrum. The C line is almost always called Hα (the level 2–3 transition of hydrogen), while the H&K lines are still called by that designation and are resonance (ground state) lines of ionized calcium. The formation of these lines is discussed in Appendix A and their use as diagnostics are explained in Section 3.2. Although the gas pressure dominates the magnetic pressure in the photosphere (except in sunspots), the reverse becomes true in the upper layers of the solar atmosphere. This leads to the atmosphere being structured along magnetic field lines since the gas is essentially a fully ionized plasma. Images of the Sun made in these diagnostics show a wonderful array of structures of many different sizes. There are small bright points, large patches of brightness called "active regions," small dynamic linear features called fibrils and spicules, and the larger filaments and prominences (Chapter 3).

Because both the required opacities (given the lower densities) and temperatures are greater in the chromosphere, the UV (starting below about 330 nm where the Earth's atmosphere is opaque) and FUV (between about 170 nm and the hydrogen Lyman continuum) is a better part of the spectrum in which to observe it. Unfortunately, this means the observations must be gathered outside the Earth's atmosphere. Finally, the corona is far more transparent and much hotter, so the EUV (below 91.2 nm where the hydrogen Lyman continuum starts) and X-ray parts of the spectrum are needed for directly observing coronal structures. Figure 1.1 shows the Sun at the same time in a variety of diagnostics at different wavelengths that sample different parts of the solar atmosphere. As mentioned above, visible light from electron scattering does illuminate parts of the coronal structure and make visible coronal mass ejections, but it is not diagnostic of the main sites of coronal heating.

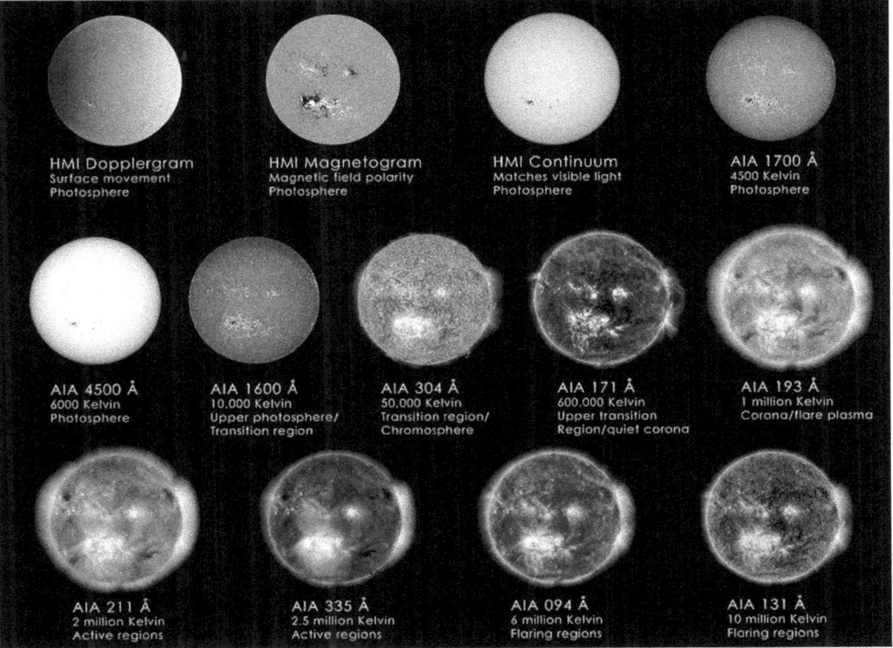

Figure 1.1. Views of the Sun on the same day at many wavelengths. The images were taken by the Solar Dynamics Observatory operated by NASA. They show the appearance of the Sun at different layers moving steadily upward to higher temperatures. The first two (upper left) depict the velocity field and magnetic fields in the photosphere, the next is the photosphere itself, followed by the temperature minimum, and upper photosphere. The 1600 Å image shows the chromosphere, the 30 Å image shows the base of the transition region, the 171 Å image is in the transition region, and above that we are moving higher in the corona. There is a clear correspondence of the brighter regions with stronger magnetic fields. Credit: Reprinted with permission from NASA/SDO. For a nice video presentation of the Sun at different wavelengths displaying amazing activity, try https://svs.gsfc.nasa.gov/cgi-bin/details.cgi?aid=11742. Credit: NASA/SDO, Goddard Space Flight Center.

Solar flares are even hotter at their main heating sites and so require even shorter wavelength observations to see their direct manifestations. They do indirectly excite visible emission through the impact of particles on the photosphere and chromosphere, especially Hα emission. X-rays (and even more so, γ-rays) are hard to make high-resolution images with (reflection optics no longer are effective), so only in the past couple decades has it been possible to make images from space with resolution similar to optical images of the Sun. The fact that particles are being accelerated along magnetic fields also means that radio emission is produced and can be dynamically observed in intensity, spectrally, and imaged with interferometers. This topic is covered in Section 4.3.

Introductory astronomy books cover the basics of the solar cycle well. The source of the magnetic field in the current Sun is a magnetic dynamo. Current research on dynamos is discussed in Section 6.1; here it suffices to mention that the solar dynamo is a cyclic internal mechanism of generating structured magnetic fields, which then dissipate but are regenerated. During an activity cycle the Sun starts in a relatively inactive state, with few to no spots visible and the corona configured with relatively more open field regions which lead to coronal holes near the surface and fewer and less symmetric extended streamers. As the cycle progresses, more sunspots appear at mid-latitudes (30–40°) in both hemispheres with a preferred orientation of the commonly bipolar pairs. Their orientation slowly tilts and the preferred latitude migrates toward the equator as the spot number increases. Along with the sunspot activity, other forms of magnetic activity (total magnetic flux, UV and X-ray brightness, number of active regions, flare rates) also wax and wane, and the structure of the corona becomes less open and more concentrated into closed active regions with increased flaring as the spot number increases.

As the spots get to low latitudes after about 11 years their numbers decrease, and that part of the cycle fades. The second part of the full 22 year cycle starts a year or two before the first part has fully disappeared; activity is again at a minimum. It unfolds with similar behavior but with the opposite polarity for both the large-scale dipole and the emerging small bipolar regions. The maximum of the sunspot number varies from 11 year activity cycle to cycle (this is usually what is meant by "sunspot cycle" or solar cycle) with about a factor of 3 between the weakest and strongest well-observed cycles. There are also occasionally cycles with almost no sunspots, called "Maunder Minima." Cycles are covered in more detail in Section 6.1.1.

In addition to the cyclic large-scale field, there is a constant presence of smaller-scale fields. These manifest as the chromospheric "network" (often observed using the H&K lines) which traces the boundaries of convective supergranules which sweep the magnetic field to their edges (the faint filigree away from active regions in Figure 3.2). This component of the field can also be seen as tiny bright points in very high-resolution images. It presumably arises from a more turbulent and smaller-scale form of dynamo, which is not cyclic. This type of dynamo probably operates throughout the convection zone, while the cyclic dynamo is thought to operate in part through dynamics at the bottom of the convection zone where there is shear with the radiative core below (the "tachocline"). This is discussed in more detail in Section 6.1. The direct measurement of magnetic fields on other stars is possible through spectroscopy and polarization measurements (albeit rather difficult ones). Some types of observations provide only an average field and filling factor, but other types contain information on the spatial distribution of the fields. These techniques are the subject of Section 6.2.

The Sun is by far the most well-studied case of stellar magnetic activity. As one looks to other stars, there is very little opportunity for spatial resolution, the strength of the signals is much weaker, and the opportunity for nearly constant time coverage is greatly reduced. On the other hand, the Sun provides us with only a single example of stellar mass, age, rotation rate, convective state, and composition. It is therefore essential that we do our best to learn what we can from a wide variety of other stars, observed in as many diagnostics and methodologies as possible. Only then will we understand the general phenomena of stellar magnetic activity, and their effects on both the stars and their possible planetary companions. Some of what we have learned from other stars is discussed in Chapters 5, 6, and 7.

An Introduction to Stellar Magnetic Activity

Gibor Basri

Chapter 2

Photospheres

One of the simplest measurements one can make (at least conceptually) is of the overall brightness of a star in some broadband portion of the spectrum, like the visual range. In practice this is not as easy as it sounds, particularly if one wants sufficient precision to gain valuable information from the visible intensity variations of a star like the Sun. Such a task requires photometry with a precision of a few parts per hundred thousand. Even for a star as bright as the Sun this is difficult from the ground, and even from space it is subject to subtle instrumental issues. Other stars of course are vastly fainter, but that can be addressed in principle by increasing the collecting area of the telescope. Also of interest are variations of integrated intensities in wavelength bands both shortward and longward of the visible, and the variation of the total bolometric luminosity.

There are three primary causes of stellar photometric variability. One is the possible pulsations and vibrations of the stellar surface. Although solar-type stars are not typically in the pulsational "instability strip," they still experience a variety of non-radial pulsations. Caused by the "noise" due to convection and the various types of resonant cavities produced by the stellar structure, these very subtle variations contain information on the interior structure of the star, leading to the burgeoning field of "asteroseismology." From the five-minute solar "p-mode" oscillations (one of the most well-known and obvious modes) to a large variety of internal pressure modes, they have given us a fairly accurate view into the interior structure of the Sun. They are usually measured with spatial resolution on the disk of the Sun but can also be seen in total disk-integrated light. With the advent of precision photometry space missions (MOST; COROT; Kepler; TESS; and more to follow), the field of asteroseismology has entered a golden era. I do not discuss this rich topic further, however, because it is peripheral to our concentration on magnetic activity. An excellent recent review of this field can be found in Aerts (2021).

The next photometric variation we discuss is in the absolute irradiance of a star (either in a wavelength band or bolometric). The full bolometric variation of the

Sun is often designated TSI for "total solar irradiance"; it is measured from space at 1 au. I will use it more generally to mean the stellar bolometric luminosity. This is a difficult measurement because it requires absolute accuracy in addition to a level of precision of better than a part in ten thousand. The measurement of most interest here is the variation in TSI caused by the presence of magnetic cycles (not the very slow increase of stellar luminosity due to stellar evolution). This is related to the sunspot activity cycle which changes over about a decade on the Sun, so the measurements must be absolutely accurate (or at least differentially calibrated) to a very precise level over a long period of time. This is separate from the monthly differential variations caused by the presence of individual spots or spot groups (those are discussed below). It refers to the change in TSI averaged over a number of stellar rotations at different phases of the magnetic cycle. Such changes can arise due to the average number of spots present (and their locations) along with the average integrated area of faculae (which may or may not be accompanied by spots).

There is a substantial literature concerning the TSI of the Sun. It has been measured from space for several decades now by different instruments (Figure 2.1). There has been an ongoing effort to reconcile the calibrations of these different missions. One review of these efforts appears in Yeo et al. (2014). It is clear that the TSI increases along with the sunspot number, despite the naive expectation that having more dark spots on the surface should lead to lower intensity. The reason for this is that the faculae (which are bright) increase in area and persistence more than the spots. Also relevant is the fact that spots have their maximal effect near disk center (where they have the largest projected area to an observer), while faculae are brightest near the limb (for geometrical reasons having to do with the mechanism that produces them) and cover more area. The end result is that the faculae drive the

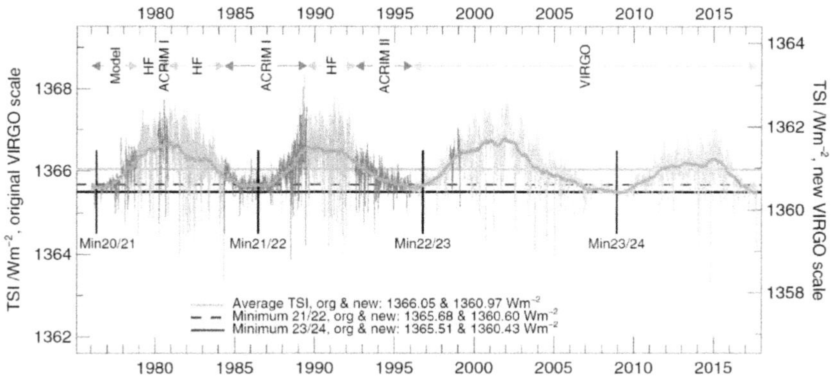

Figure 2.1. The re-calibrated total solar irradiance (TSI) measured by several spacecraft instruments (listed above the curve) over most of Cycles 21–24 (cf Yeo et al. 2014). The "noise" in the colored lines is really signal; the downward spikes are caused by the passage of spot groups across disk center. It is clear, however, that the TSI is actually larger on average (gray line) when more spots are present. Credit: Reprinted with permission from Wolfgang Finsterle, © PMOD/WRC.

integrated (both over the disk and over several rotations) total output, so that stronger magnetic activity leads to an increased total luminosity. This behavior has been well modeled for the Sun by Shapiro et al. (2017).

The situation for other stars is more complicated. Two distinct types of behavior are seen (Lockwood et al. 2007; Hall et al. 2007; Hall, 2009). Many stars follow the behavior of the Sun in brightening when spots are more prevalent. But sufficiently chromospherically active stars (which have larger surface magnetic fluxes) show the opposite behavior: the overall brightness of the star is lower when starspots are more prevalent. This means that the presence of large or numerous enough spots overwhelms the presence of faculae. A detailed analysis of how this happens and under what conditions has been provided by Shapiro et al. (2016).

There is a more diverse set of behaviors if one considers a select wavelength band rather than the bolometric luminosity. Because the chromosphere and corona are hotter their thermal emission contributes more to shorter wavelengths. In addition their filling factors are more variable so there is a greater contrast between the quiet and active Sun at higher energies. The variation in luminosity is increasingly extreme as one moves to shorter wavelengths; the Sun can be several times brighter in X-rays at solar maximum compared to solar minimum. The youngest and most active stars can be up to 1000 times more luminous in X-rays as a fraction of their bolometric luminosity (10^{-3}) than the Sun is (10^{-6}).

2.1 Physical Structures in Photospheres

Magnetic fields occur on the Sun with a large range of scales. With the very high angular resolution afforded by adaptive optics on large telescopes (like DKIST), flux tubes can be seen almost down to the limiting photon mean-free-path below which they would not produce a visible structure. Magnetic structure also exists up to the scale of a global dipole field. This is a span of scales between 50 and 500,000 km. Whether features appear as bright or dark (or at all) in a visible image depends in part on the temperature profile and contrast between the magnetic region and non-magnetic surroundings, and in part on their relative opacity and geometrical configurations. Furthermore, they are dynamic phenomena so they vary on a variety of timescales. Part of this variation is due to magnetic flux emergence and dissipation, and part of it is due to gas flows moving flux around and causing it to be concentrated, dispersed, or dissipated, and/or to interact with other flux of the same or opposite polarity.

In and below the photosphere the temperature is expected to decrease toward the outside because it is in principle a radiative equilibrium structure (cf Appendix A). Flows below the stellar surface tend to concentrate the magnetic field first into small flux tubes with fields ranging from 1000–1500 Gauss. One might expect the temperature difference inside and outside a small flux tube to be minimal, since photons can easily diffuse through the sides (this is what is meant by "small"). Nonetheless, if there is a reasonably strong field in the tube it will supply some of the total pressure, which must balance the outside pressure or the tube will contract or expand. Since there is both magnetic and gas pressure inside the tube while there is only gas pressure outside the tube, there will be less gas inside the tube. The diminished amount of gas means that the opacity per unit length inside the tube is less than outside the tube. If an observer is looking straight down into a vertical tube then they should see the tube as a bright feature relative to the surrounding photosphere because the lower opacity in the tube means one can see to a greater depth. If the temperature profile is the same inside and out, this greater depth will be hotter and brighter than without the field.

If the size of the tube is increased various effects can occur. One possibility is that the tube acts as a "light leak" in the photosphere since it has lower opacity. Energy from the sides has an easier time escaping than it would without the tube, and one might therefore get some cooling and develop a dimming ring around the tube. Tilting the tube or looking in at an angle will of course alter what one sees since now one is looking partially at the sides instead of the bottom and the slant path of opacity is different. If the tube is large enough and its field strong enough, convection can begin to be inhibited inside the tube which will cause cooling. The first really darker features seen on the Sun are "pores," which look like small beginnings of sunspots. They have internal fields of a little over 2000 Gauss and variable sizes from one to a few thousand km across. Figure 2.2 is a very high resolution image of the surface of the Sun showing some of these features. It is worth mentioning that these features are only dark by comparison with the solar photosphere; they look more like the photosphere on a K or M star.

As larger flux concentrations appear they begin to develop darker cores (umbrae) and less dark surrounding fingers (penumbrae)—a sunspot (Figure 2.3).

Figure 2.2. A high angular resolution image of a portion of an active region near the limb, taken by the Swedish Solar Telescope using adaptive optics in the continuum at 487.7 nm (tics are 1″). The pervasive cells (1000 km wide) are the convective granulation pattern. Some of them have become faculae, appearing as small bright "cliffs" on the sides of the granules facing the observer. The image also contains pores and small sunspots. Credit: Reprinted with permission from the Institute of Solar Physics. Image observed with the Swedish 1-m Solar Telescope. Observations: Göran Scharmer, ISP. Image processing: Mats Löfdahl, ISP.

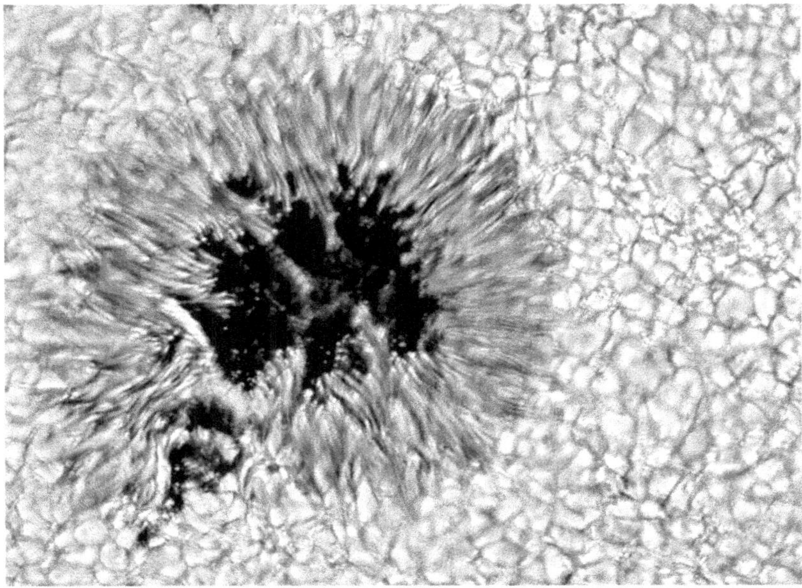

Figure 2.3. A high angular resolution image of the solar photosphere taken by the National Solar Observatory vacuum telescope. It shows a sunspot with the dark umbrae and filamentary penumbra. Granules are visible outside the spot, disturbed to various degrees by the magnetic field. Vertical magnetic flux tubes swept into the intergranular lanes are visible as tiny bright points (100 km wide) right of the spot. Credit: Reprinted with permission from NOIRLab. Credit: T. Rimmele (NSO), M. Hanna (NOAO)/AURA/NSF.

The rising large flux tube expands above the photosphere; it is usually connected to another tube of the opposite polarity that appears at the same time in a grand Ω configuration. Solar umbrae can range in size from 3000–30,000 km although as they get large they tend to have complex non-circular structures. Their internal fields are close to vertical with strengths that range as high as over 3000 Gauss. Their core temperatures can drop below 4000 K compared with the temperature of the quiet photosphere of around 5800 K. Because the optical depth scale is different at umbral temperatures, the "surface" is about 500 km lower than in the quiet Sun. This is called the "Wilson depression." Penumbrae have fields a few hundred Gauss weaker and the field is oriented more close to horizontally. True sunspots usually emerge in bipolar pairs, as a large underlying flux loop which must connect with itself below pierces the photosphere at two locations separated by the emerging diameter of the loop. There is a large literature describing the polarity, separation, location in latitude, tilt with respect to the pole, and motion of these spot pairs, which reverse their properties in the two halves of the 22 year solar magnetic cycle (cf Section 6.1.1).

When looking at lower resolution the granulation pattern appears as a "graininess" to the Sun (Figure 2.4). Flux emergence also happens on scales that do not produce pores or spots. Flux arrives at the surface in small bipolar configurations consisting of many tiny flux tubes of a few hundred Gauss. Convective motions push

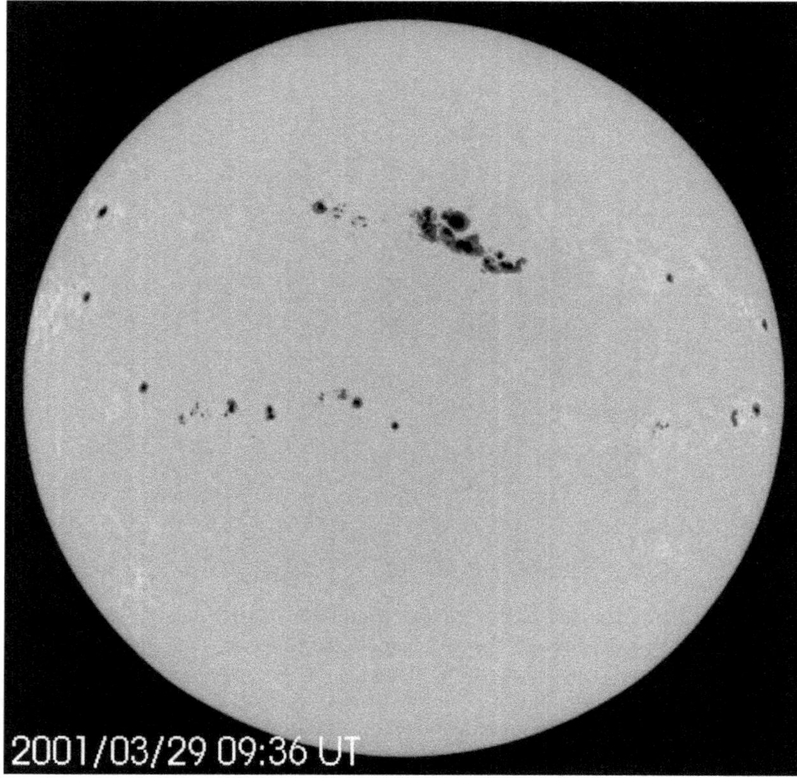

Figure 2.4. A full disk visible image of the very active Sun taken by NASA's Solar and Heliospheric Observatory (SOHO) spacecraft. It shows one of the most spotted days on the Sun in recent history. Notice that faculae are subtly visible as brighter (yellow in this image) patches associated with the spotted active regions near both limbs as well as an unspotted one to the lower left, but are not visible in the even more active region near disk center. Credit: Courtesy of SOHO/MDI consortium. SOHO is a project of international cooperation between ESA and NASA. For a video of the Sun rotating several times and sunspots coming and going, see https://sohowww.nascom.nasa.gov/bestofsoho/Movies/10th/MDI-Sunspot-Crossings-2001-large.mpg. Credit: SOHO/LASCO (ESA & NASA)—Brendan Gallagher (NRL).

them together and also cause "convective collapse" that concentrates the field to above 1000 Gauss. This happens when the bipolar field rises above the photosphere (where magnetic pressure dominates) and forms a small Ω-shaped configuration, allowing gas to drain down the side tubes and the photospheric pressure to compress the field. The tubes organize themselves into more coherent bipolar regions in a day or two as an active region emerges.

When there is enough of such field, the phenomenon of faculae emerges. This is an effect related to the brightness of small vertical flux tubes. The field swept to the sides of granules reduces the opacity there, allowing one to see further into the hotter core of the granule. The effect is much easier to see when looking in from the side than looking down from on top (it is relatively rarer to see bright bottoms of flux tubes since they must be straight and vertical). Very high spatial resolution observations closer to the limb show the granules with bright "cliffs" and darker centers (Figure 2.2). An excellent set of images and discussion of models of faculae can be found in De Pontieu et al. (2006).

Broader concentrations of magnetic field are called "active regions." In visible light they will contain collections of faculae only visible near the limb of the disk (Figure 2.4). They also often contain pores or sunspots that can be seen in visible light anywhere on the disk, although they are harder to see near the limb due to area projection effects. Active regions are visible at any angle in magnetograms, chromospheric, and coronal diagnostics (Figure 1.1). Because of their greater areal coverage the effect of faculae on the total light curve is a net brightening when active regions are closer to the limb, while spots produce a darkening primarily when they are near disk center. As mentioned above, the TSI measurements of the Sun show that the faculae produce more brightening than the spots produce darkening on a timescale of order a fraction of the cycle, but at a given moment the Sun may be darker than average if there are a number of spots near disk center.

Finally there is also a spatial inhomogeneity in the other direction: exceptionally cool patches of photosphere. These are observed as portions of the solar photosphere that unexpectedly show strong CO absorption, but they have no significant effect on the visible appearance of the surface. The CO molecule should not form at the temperatures of the radiative equilibrium solar photosphere, so something is causing portions of the atmosphere to cool to temperatures (3500 K) well below the classic temperature minimum (4200 K) and extend into the middle chromosphere where temperatures are above 7000 K. These structures have been dubbed the "COmosphere" by their primary advocate (Ayres 2002). There have been fierce arguments about how common or stable these structures are, but they are undeniably present in high resolution infrared spectra of the Sun. Their exact nature and the mechanism that produces them remains to be understood; some kind of thermal instability is likely since the CO molecule is a very efficient coolant. The point is that atmospheric studies that assume horizontal homogeneity whether in the photosphere or above will fail at some level of precision because stars simply do not have that characteristic.

The sunspot cycle (which is also an active region cycle) is associated with the emergence of fields from a large-scale internal magnetic dynamo. It is quite obvious in Figure 2.1. There is also a more or less constant emergence of small-scale flux all over the Sun. This is thought to be related to a more localized turbulent dynamo mode. This small-scale field tends to be swept in aggregate to the boundaries of the supergranulation convection pattern, producing what is called the "network" that can be seen in magnetograms or diagnostics like Ca II (cf Chapter 3). A much more detailed discussion of the topics in this section can be found in the book "Solar and Stellar Magnetic Activity" (Schrijver & Zwaan 2008; chapters 4, 5) and in the book "The Sun as a Guide to Stellar Physics" (Engvold et al. 2019).

2.2 Starspot Light Curves

The most obvious and easily measured photometric variation caused by stellar magnetic fields is the effect of starspots on a precision light curve. Because spots are cooler than the surrounding photosphere, they reduce the total luminosity of the visible hemisphere. The temperature differential between the spot and quiet star is a (not very well determined) function of the stellar effective temperature, with a difference of order 2000 K for solar-type stars and perhaps a few hundred degrees for M dwarfs (whose effective temperatures are similar to umbral spot temperatures on the Sun). The drop in flux in a light curve at a given moment depends on the number, size, darkness, and limb position of all the spots on the currently visible hemisphere. Unfortunately, the fact that all these factors go into determining that single number means that the information inherent in a precision light curve is quite degenerate over those parameters, limiting what can actually be learned about the physical nature and distribution of spots from just a light curve.

Until about 2005 it was difficult to measure light curves to a precision of a milli-magnitude, about 0.25% or 2.5 parts per thousand (ppt). This is roughly the maximum visible photometric variation exhibited by the Sun during solar maximum. The ability to monitor stars with a high observing cadence was also quite limited because it was mostly carried out from ground-based observatories. To monitor more than a few stars for lengthy periods requires robotic telescopes. Any such project is subject to the steady interruptions by daylight as well as interruptions due to weather, and the fact that stars can be largely unavailable for part of the year when brought too near the Sun by the Earth's orbit.

With the advent of space-based photometric missions the coverage situation greatly improved along with the photometric precision. Missions such as MOST, COROT, Kepler (and K2), and TESS have revolutionized the study of stellar light curves, with more missions planned for the future. Observing from space means an almost continuous duty cycle (while the telescope remains pointed at the star) and precision down to less than 0.1 ppt in the best cases. The photometry is often differential rather than absolute, but that is sufficient to see at least the basic signature of starspots covering varying amounts of the visible stellar hemisphere. They vary simply because they are on a rotating star but also grow and decay on the stellar surface with some timescale. The amount of photometric variability due to starspots on a star can be characterized in a very basic way by the maximum amplitude of the differential light curve over many rotations of the star (the "range").

The Kepler mission obtained such data for four years continuously on roughly 200,000 stars in a field of view over 100 square degrees in an area chosen a little above the Galactic plane. The stars sampled include a segment of the solar neighborhood out to about a kiloparsec that is a reasonably representative sample of the Milky Way outside the bulge. The sample was intentionally strongly biased toward solar-type stars (effective temperatures between 5000–6500 K) but included stars down to about half a solar mass (early M spectral types 3300–3800 K). In such a sample one might guess that about two-thirds of the stars are the age of the Sun or older, since the Sun is one-third the age of the Galaxy. Older stars should be about as active as the Sun or less active (Section 5.3). A caveat is that older stars are kinematically heated into a scale height more than a kiloparsec, so the Kepler sample is likely biased toward thin-disk younger stars. Galactic models have been applied to this question more accurately now that Gaia data is available (van Saders et al. 2019).

The Kepler photometer was able to detect solar levels of starspot activity on stars brighter than about Kepler magnitude 13.5 but included fainter stars to about magnitude 16. Thus there is a significant fraction (more than half) of the Kepler solar-type stars that show no clear starspot signal. Examples of three light curves from Kepler with different levels of variability are shown in Figure 2.5. The bulk of photometric ranges for spectral types between F and K fit within the range of the solar activity cycle, although the Kepler distribution is concentrated somewhat toward the upper end. In addition, about a quarter of the solar-type stars have higher ranges than the Sun ever exhibits (Basri et al. 2013). These are presumably the younger stars (Section 5.2), and indeed many of them have shorter rotation periods than the Sun. There is also a clear trend toward larger photometric ranges as the stellar temperature decreases below 5000 K. Although the sample size decreases with the temperature due to the intrinsic drop in stellar luminosity (Kepler's magnitude-limited sample comes from smaller volumes) it is clear that the lower bound on the photometric variability range is rising. By the time we reach 4000 K, there are

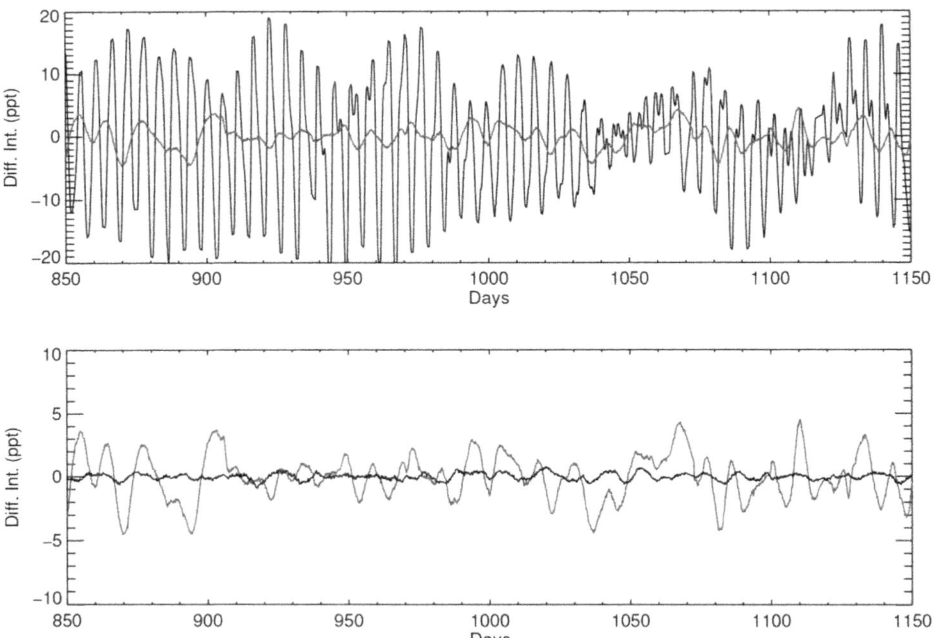

Figure 2.5. Portions of the differential light curves of 3 stars collected by the Kepler mission. They are shown in parts per thousand in 5 hour bins and median subtracted. The top panel has a star much more rapidly rotating and active than the Sun (black) and one somewhat more rapid and active (blue). The lower panel repeats the upper blue curve and has a star of comparable activity to the Sun in black. The KIC indices of the three stars are (in order of decreasing variability) 4449749, 5089026, and 8313989. Their effective temperatures are 5980 K, 5590 K, and 5860 K, and their rotation periods from McQuillan et al. (2014) are 7.0, 20.8, and 39.4 days.

essentially no stars whose variability is less than the average solar range and most of them are more variable than the solar maximum.

It is easier to see chromospheric plasma against cooler photospheres because of the temperature/wavelength sensitivity of the Planck function, but that does not necessarily explain why variability from a cooler star should be larger. The temperature contrast between cool spots and cooler photospheres is expected to be fractionally lower. Photometric variations due to convective cells might be expected to be lower as well due to the fact that there are more and smaller cells on stars with higher surface gravity. Facular models also predict smaller contrast on such stars (Beeck et al. 2013). One mechanism that works in favor of higher broadband variability in cooler stars is the fact that a fixed temperature difference in Kelvins will produce a larger contrast at visible wavelengths due to the fact the visible is in the Wien part of the blackbody spectrum for temperatures significantly cooler than solar. Perhaps there is also enough microflaring or some other process that produces variable hotter regions above these cool photospheres. This question awaits a solid explanation.

The light curves seen by Kepler fall into three basic categories. There is a large set of essentially non-variable stars (except for noise). These are presumably older stars that no longer have significant spot groups on them (at least most of the time), although a few could be stars experiencing a "Grand Minimum" (Section 6.1.1). There is a set of stars that show clear stellar photometric variability, but it is fairly disorganized and hard to measure a periodicity from. The Sun falls into this group; solar light curves are resistant to yielding the solar rotation period most of the time. The lowest range curve in Figure 2.5 is an example of this; the 40-day period from McQuillan et al. (2014) is incompatible with the duration of some of its dips unless they are due spot evolution rather than stellar rotation. If spots (by which I usually mean "spot groups") significantly change their size while visible during one rotation, that can causes changes in the intensity that are not primarily due to rotation. This type of behavior seems to be associated with stars in the age range of the Sun. About a third of the Kepler stars belong to the third set. These have larger than solar ranges with clear periodicities, and increasingly organized pattern changes as the range increases. The higher range curves in Figure 2.5 are examples from this set. They tend to have rotation periods (and ages) that are shorter than for the Sun. The spot group lifetimes (in rotation periods) also seem to get longer on more rapid rotators so their effects last for an increasing number of rotations and longer term patterns appear (eg. the largest range curve in Figure 2.5). This reaches an extreme in very young stars, where the photometric variations can look the same for tens or even hundreds of rotations.

Gilliland et al. (2011) claimed that the Sun is unusually quiet, based on the large sample of solar-type stars from Kepler. This was refuted by Basri et al. (2013); who found that the variability measure in the previous paper is unsuitable for answering that question. They show that the Sun very easily fits within the Kepler solar-type sample photometrically. This is true whether talking about differential variability on timescales of hours up to weeks. The absolute variability including faculae was not measured. More recently Reinhold et al. (2020) pointed out that there are some stars with nearly the Sun's rotation period and temperature that are indeed more photometrically variable. The question is how common that is, since many solar-type stars do not have measured periods in the Kepler sample and the rotational sample is biased toward greater photometric variability. The correct interpretation is likely that there is a small fraction of stars like the Sun that show more persistent (longer-lived) spot groupings than it does. These lead to larger photometric amplitudes and the ability to detect their solar-like rotation periods, whereas the Sun's period is typically undetectable because its spot groups are too short-lived. This is discussed further at the end of this section.

One of the most obvious properties of the organized light curves is that they tend to show either one or two dips in intensity per stellar rotation. The behavior of the light curves is much simpler than the behavior of the underlying starspot distributions, primarily because the light curves sample the entire visible hemisphere at any given time. Depending on the starspot distribution (ignoring time evolution) one can often divide the star into a brighter and darker hemisphere, despite the presence of spots of different sizes scattered around in various ways. In that case, the light curve will exhibit a single dip over one rotation; the minimum indicates when the darker hemisphere is most facing the observer. If the spot distribution is complex enough in a broadly distributed way, the visible hemisphere may darken and brighten more than once during a rotation. It is hard for that to happen more than twice if spots typically last more than one rotation, given the very low spatial resolution of the light curve. In that case one usually gets two dips per rotation, and their relative depths can be similar or different. Basri & Nguyen (2018) found that the fraction of a light curve that is double-dipped increases with the rotation period.

The most variable light curve in Figure 2.5 shows these characteristics, having mostly single dips, but exhibiting a period of doubling after about Day 1040. The double dips are obvious because one is much smaller than the other. It is less obvious that most of the dips in the intermediate (blue) curve are double unless one checks the rotation period against their durations. Indeed, it is not obvious that the rotation period isn't just half what is stated. The inferred period comes from the overall behavior of the light curve, not just from this segment. This possible confusion is a constant issue when trying to determine rotation periods from light curves. The two curves with smaller range also contain a few small dips that are implausibly short and likely to be instrumental effects or signatures of fast spot evolution.

Starspots vary their photometric influence either because they are physically growing or shrinking or because their projected area is changing as they pass closer to or further away from the sub-observer point. Their distribution on the stellar surface can also change because of either starspot evolution or differential rotation. These changes mean that stars can transition from a single dip to double dip light curve mode on a variety of timescales, and that the double dip structure (when present) can show an evolving pattern of the placement and depth of the dips. As mentioned above this can lead to confusion between a half and whole period, especially if the light curve does not contain a reasonable number of rotations.

One hope has been to utilize these changes to infer something about spot distributions, spot lifetimes, and surface differential rotation (assumed to be latitudinal). Basri & Shah (2020) provide a detailed discussion of the difficulties this project has encountered. It is very hard to accurately and confidently measure surface differential rotation using just a light curve. The effects of spot evolution are quite similar to differential rotation unless spot lifetimes are many rotations. It does appear promising to be able to infer something about starspot (group) lifetimes since the light curves are rather different for spots that last only a few rotations compared with spots that last for many rotations. Of course, there may be spots of different sizes with different lifetimes on the same star; on the Sun small spots are more numerous but die more quickly than large spots.

It is reasonable to anticipate progress soon on this quest. New methodologies will need to be developed to make progress on that front, which is of great interest for dynamo theories of magnetic field production and cycles. A promising avenue is that if spots either live for several rotations (Basri & Shah 2020) or keep re-emerging at similar longitudes (Isık et al. 2020) the light curve will exhibit larger range and more periodic structure. A paper is about to be published by my group in 2021 that utilizes this effect as expressed by the strengths of the first few autocorrelation peaks from the light curves to infer the lifetimes of spot groups or spot emergence regions. Stars with larger variability tend to have longer spot lifetimes. We find that there is a definite structure to the lifetime-rotation diagram, with a branch of more rapid rotators with longer lifetimes and another branch with slow rotation and short spot lifetimes. Cooler stars tend to have longer lifetimes at the same rotation rate, which may be part of why they seem to be more photometrically variable.

Faculae provide much larger absolute intensity changes than spots (at least for stars like the Sun) as mentioned above. The source of these changes is the absorption line spectrum (changing line strengths) so these changes also become stronger metallicity is increased (Witzke et al. 2018). The Kepler mission, however, is much less effective at detecting faculae than starspots. The facular signal does not tend to produce the sort of sharp features that spot groups do. The primary indication of faculae is found in changes in the absolute flux from a star but Kepler only measures changes in differential flux. This issue is discussed in detail by Basri (2018) who shows that when solar data is treated as though observed by Kepler the facular signal disappears. The exception to this is during times when spots are essentially absent; there is then a small smooth variation over a rotation due to faculae but such signals are essentially absent in Kepler data. There is an exception to the statement that Kepler does not measure absolute fluxes: it did so every time there was a data dump to the Earth or when quarterly spacecraft rotations took place. In these instances full-frame CCD images were downloaded, and those provide an opportunity to measure the absolute brightness of a target star against an average of many stars nearby on the same detector. Montet et al. (2017) took advantage of this to produce absolute flux curves from Kepler and claimed facular detections, but Basri (2018) raised concerns about their reality.

2.3 Doppler and Transit Imaging of Starspots

The problem of mapping starspot distributions becomes more tractable with the addition of spectroscopic information. The basic reason is that because stars rotate, the approaching side will be blueshifted and the receding side will be redshifted by an amount that varies with the stellar inclination and the angle to the sub-observer point. A starspot reduces the amount of light coming from a certain area on the star with a certain Doppler shift. Paradoxically, if one is observing a Doppler-broadened spectral line the presence of a starspot at a particular velocity will produce a "bump" (slightly less absorption) at that velocity (Vogt & Penrod 1983). This is because the line will have reduced contrast at velocities that contain spots because the continuum is reduced. Utilizing this information requires high spectral resolution and high signal-to-noise since the line profile must be well-resolved to separate out Doppler slices and the starspot signatures can be small compared to the line depth. It is best to use spectral lines that are relatively isolated from other lines in wavelength and not too temperature-dependent (if the line is much stronger in a spot than the photosphere that reduces the signal).

The ideal case is an intermediate stellar inclination so that spots remain in view for more of a rotation and since a pole-on star will have no Doppler shift from rotation. One can make models based purely on the brightness variations in the line, or model more carefully the atmospheres in and out of spots. In some cases one infers bright as well as dark spots (this is methodology-dependent). One can sometimes use autocorrelation techniques to enhance the signal from many lines. Some form of regularization (like maximum entropy) must be employed because somewhat different spot distributions can produce the same line deformations. Doppler images tend to be "blurry" and as simple as possible because of this. Agglomerations of smaller spots that are concentrated enough will be interpreted as a larger spot. There is of course a lower limit to the diameter of a spot group that can be detected (roughly ten degrees). Models often make the assumption of fixed temperature contrast between spots and photosphere that is undoubtedly not fully true or size-independent. Figure 2.6 shows an example of Doppler images made after separating the Doppler-shifted spectral lines from each binary component in a close system (an extra wrinkle).

One result that is often found is that rapidly rotating stars have polar or high-latitude spots that can cover a substantial portion (tens of percent) of the visible surface. The spatial resolution on the star in a Doppler image is dependent on the rotation velocity (the more rapid the rotation the better the separation of Doppler slices). On the other hand, the nearer the pole a spot is the smaller will be its Doppler shifts. Questions have been raised about whether the inference of a polar spot is simply an error in the interpretation of the depth of the center of the line (since polar spots will tend to sit near line center). This can sometimes be settled by studying the color of the star (spots are cooler and redder) or the behavior of temperature-sensitive spectral lines (see Section 2.4) that look different in spots. It is now fairly confidently established that rapid rotators do indeed have a lot of high-latitude spottedness. There are theoretical reasons to support this possibility because rapid rotation can mean that Coriolis forces overwhelm buoyancy forces and flux tubes rise from the deep interior along cylinders (Schuessler & Solanki 1992; Isık 2018). A milder form of this mechanism for convection is responsible for surface differential rotation like that seen on the Sun.

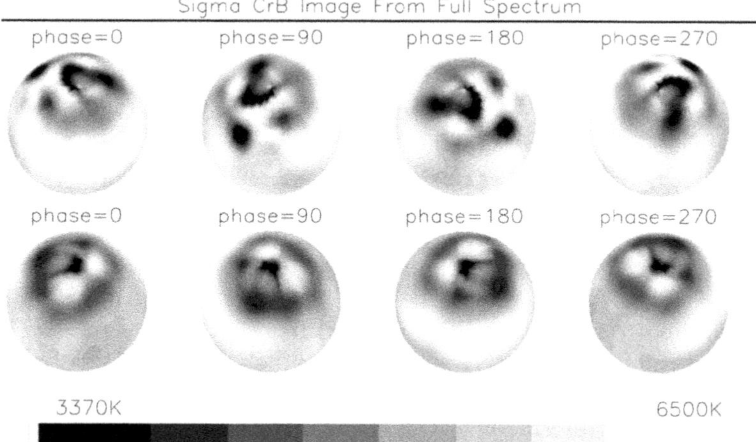

Figure 2.6. Doppler images of starspots on the young close double-lined binary system σ CrB. Both components (each row) are slightly hotter than the Sun and young. Several lines were employed with a regularization scheme for each image. Credit: Reproduced with permission from Astronomy & Astrophysics, Strassmeier & Rice (2003) © ESO.

In principle, amassing a long time series of Doppler images can be very helpful to a variety of topics. One might be able to see surface differential rotation (if the pattern changes slowly enough and features remain recognizable as they drift). One could detect the the presence of a cycle and possibly see the migration of magnetic field poleward or equatorward during a cycle. One can see whether large spot configurations are collecting or dispersing. One might get an idea of emergence, decay, and persistence timescales (or at least configuration lifetimes) and how they depend on spot coverage or stellar rotation. All of these ideas have been tested to some level (typically on very active stars). Strassmeier (2009) provides a review of results from Doppler imaging.

There is another way to "directly" image starspots, and that is to employ exoplanet transits across their host stars. If the planet crosses over a spot it creates a "bump" in the light curve in much the same way that bumps are created in line profiles for Doppler imaging. While over the spot the planet is blocking less light than if it were over the equivalent area of photosphere. With the advent of precision photometry space missions this technique became possible, and with thousands of transiting exoplanets found there have been many measurements of spots. An example of multiple spot crossings for a particular star is shown in Figure 2.7. When combined with information about the transit chord, and even better the Rossiter–McLaughlin effect (which requires transit spectroscopy) to yield orbital obliquity, one can infer the latitude, longitude, and size of the spot or spot group. Multiple crossings provide information about spot lifetime, whether there are preferred spot latitudes and/or longitudes, and how those drift. Of interest is how the spots very explicitly observed in this way compare with spot distributions derived from light

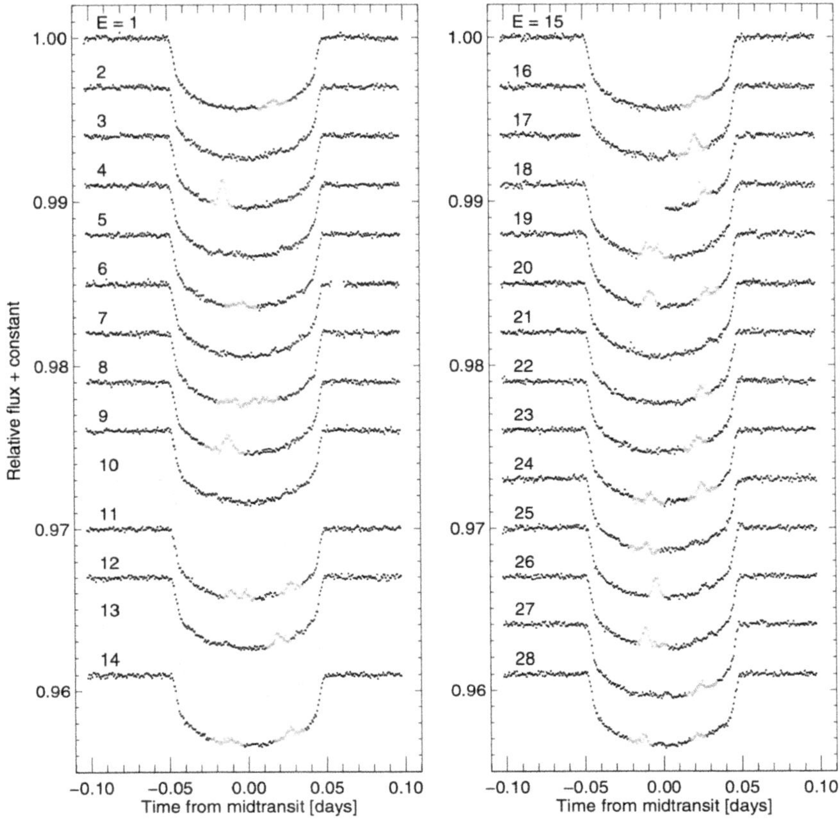

Figure 2.7. Kepler observations of HAT-P-11, a K star with a $5R_\oplus$ planet in a 5 day orbit. The light gray lines are best fit models to each light curve without the spot, and the spot features are colored red. Successive transits are numbered. Credit: Reproduced from Sanchis-Ojeda & Winn (2011). © 2011. The American Astronomical Society. All rights reserved.

curves or Doppler images. An example of a very illuminating such comparison can be found in Namekata et al. (2020).

2.4 Spectroscopic Signatures of Starspots

Another way to detect the presence of starspots, mostly the spot coverage and temperature contrast rather than the spot distribution, is to measure the effect on the spectrum of parts of a stellar atmosphere that are significantly different than the quiet photosphere. The temperature differential of spots with photospheres has been found (by this method) to be larger for hotter stars. On the Sun, for example, the photosphere is 5800 K while the interior of a large umbra lies somewhere in a range of several hundred degrees around 3500 K. This difference of around 2000 K seems to hold for solar-type stars but as one moves to cooler stars it decreases to a net difference of only a few hundred degrees for M dwarfs. When the photosphere is too hot for a molecular spectrum to form but the spot produces a molecular spectrum it can be possible to discern the fraction of the star that is covered by spots. The unfortunate catch is that because spots are cooler and darker they will have less of an effect on the total flux from the star, so the spot coverage has to be substantial for this method to work.

Morris et al. (2019) provide one recent example of this type of analysis for an active star. The star is presumed to produce two separate spectral outputs, one from the quiet photosphere and one from spots. The covering fraction and temperature difference of the two components can be fit as part of an MCMC parameter search or some other scheme for determining them. The data required are high resolution spectra that cover the range where a molecular band (like from TiO) will occur; it is also preferable that the photospheric spectrum not be too complex by itself. Figure 2.8 provides an illustration of the type of data needed and the resulting fits. That paper also provides other chromospheric diagnostics as a way of comparing spot coverage with other signs of magnetic activity. As expected, the detected filling fractions of spots are much higher than on the Sun; they have to be 100 times greater to even be seen (3% instead of 0.03%). Some stars appear to have spot filling factors greater than 50%, which brings up the question: which component is the "real" star? It is probably best just to say that such stars have two important

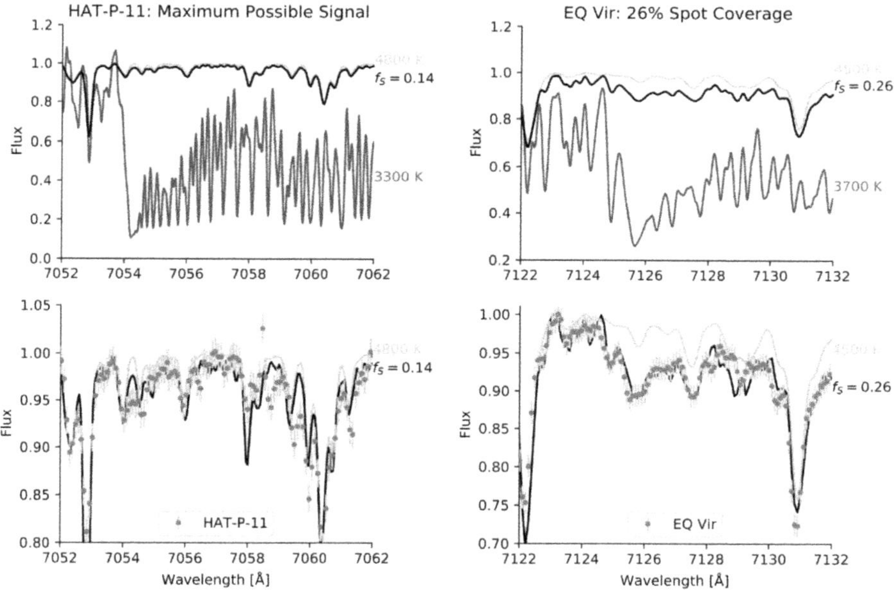

Figure 2.8. A demonstration of spectral detection of starspots using TiO bands. The upper row shows spectra from two stars (dark), models for those stars (yellow) with their temperatures and spot filling factors labeled, and models for a spot spectrum of a lower temperature (brown). The left column shows an upper limit for HAT-P11 and the right a detection for EQ Vir. The bottom row shows an expanded view of the difference between the expected photospheric spectrum and the observed spectrum. Data points are indicated with uncertainties in the bottom row. Credit: Reproduced from Morris et al. (2019). © 2019. The American Astronomical Society. All rights reserved.

atmospheric components, as one should say a zebra has black and white stripes and not wonder what the basic underlying color of the zebra is.

Another method of measuring the temperature difference and coverage of spots is through the use of temperature-sensitive line ratios. This will work when the spots do not produce strong molecular signatures or the appropriate spectral coverage is not available. It is preferable to have line pairs that are close in wavelength and not blended with each other or other lines. Even more preferable is that the lines arise from the same element (possibly in adjacent ionization stages). Other means should be used to establish what the effective temperature of the target is, then the presence of spots can be discerned through the additional strength of lines that have low excitation potentials and will be stronger in cooler gas. Catalano et al. (2002) provide an example of this technique. Because they study RS CVn systems (Section 7.3), they can check their results against Doppler imaging. They find spot temperature differences of a little under 1000 degrees and large covering fractions.

References

Aerts, C 2021, RvMP, 93, 015001

Ayres, T. R. 2002, ApJ, 575, 1104

Basri, G. 2018, ApJ, 865, 142

Basri, G., & Nguyen, H. T. 2018, ApJ, 863, 190

Basri, G., & Shah, R. 2020, ApJ, 901, 14

Basri, G., Walkowicz, L. M., & Reiners, A. 2013, ApJ, 769, 37

Beeck, B., Cameron, R. H., Reiners, A., & Schüssler, M. 2013, A&A, 558, A48

Catalano, S., Biazzo, K., Frasca, A., & Marilli, E. 2002, A&A, 394, 1009

De Pontieu, B., Carlsson, M., Stein, R., et al. 2006, ApJ, 646, 1405

Engvold, O., Vial, J.-C., & Skumanich, A. 2019, The Sun as a Guide to Stellar Physics (Berlin: Springer)

Gilliland, R. L., Chaplin, W. J., Dunham, E. W., et al. 2011, ApJS, 197, 6

Hall, J. C., Henry, G. W., Lockwood, G. W., Skiff, B. A., & Saar, S. H. 2009, AJ, 138, 312

Hall, J. C., Lockwood, G. W., & Skiff, B. A. 2007, AJ, 133, 862

Işık, E., Shapiro, A. I., Solanki, S. K., & Krivova, N. A. 2020, ApJ, 901, L12

Işık, E., Solanki, S. K., Krivova, N. A., & Shapiro, A. I. 2018, A&A, 620, A177

Lockwood, G. W., Skiff, B. A., Henry, G. W., et al. 2007, ApJS, 171, 260

McQuillan, A., Mazeh, T., & Aigrain, S. 2014, ApJS, 211, 24

Montet, B. T., Tovar, G., & Foreman-Mackey, D. 2017, ApJ, 851, 116

Morris, B. M., Curtis, J. L., Sakari, C., Hawley, S. L., & Agol, E. 2019, AJ, 158, 101

Namekata, K., Davenport, J. R. A., Morris, B. M., et al. 2020, ApJ, 891, 103

Reinhold, T., Shapiro, A. I., Solanki, S. K., et al. 2020, Sci, 368, 518

Sanchis-Ojeda, R., & Winn, J. N. 2011, ApJ, 743, 61

Schrijver, C. J., & Zwaan, C. 2008, Solar and Stellar Magnetic Activity (Berlin: Springer)

Schuessler, M., & Solanki, S. K. 1992, A&A, 264, L13

Shapiro, A. I., Solanki, S. K., Krivova, N. A., et al. 2017, NatAs, 1, 612

Shapiro, A. I., Solanki, S. K., Krivova, N. A., Yeo, K. L., & Schmutz, W. K. 2016, A&A, 589, A46

Strassmeier, K. G. 2009, A&ARv, 17, 251

Strassmeier, K. G., & Rice, J. B. 2003, A&A, 399, 315
van Saders, J. L., Pinsonneault, M. H., & Barbieri, M. 2019, ApJ, 872, 128
Vogt, S. S., & Penrod, G. D. 1983, PASP, 95, 565
Witzke, V., Shapiro, A. I., Solanki, S. K., Krivova, N. A., & Schmutz, W. 2018, A&A, 619, A146
Yeo, K. L., Krivova, N. A., Solanki, S. K., & Glassmeier, K. H. 2014, A&A, 570, A85

An Introduction to Stellar Magnetic Activity

Gibor Basri

Chapter 3

Chromospheres

3.1 Physical Structures in Chromospheres

As described in Chapter 2, the photosphere has an upper boundary in convective stars that have magnetic activity because the temperature reaches a minimum value and then begins to rise again. The source of this rise is non-radiative heating, but it probably takes several forms and the question of exactly how the solar upper atmosphere is heated has been constantly present in research on stellar magnetic activity since it began without yet having been answered fully satisfactorily. We will return to heating mechanisms in Chapter 4.

In the photosphere the gas pressure (nkT) dominates the magnetic pressure ($B^2/8\pi$). This allows convection to sweep magnetic flux tubes into the intergranular lanes, and causes flux tubes to be constantly jostled and moved around by convective motions. It is probably not a coincidence that the situation reverses in the chromosphere, where the magnetic pressure begins to dominate the gas pressure. This causes the chromosphere to be much more structured in three dimensions than the photosphere, which makes images of the Sun in chromospheric lines much more interesting. An example of this complexity is seen in Figure 3.1. The spectral lines and continua that allow us to obtain such images mostly lie in the vacuum ultraviolet (UV); there are really only three excellent lines visible to ground-based telescopes. These are the previously mentioned resonance (levels 1–2) lines of ionized calcium (Ca II H&K) and the Hα transition (levels 2–3) of neutral hydrogen. The calcium lines are in the violet at 393.3 nm (K-line) and 396.9 nm; the H-line overlaps with Hε, so generally the K-line is preferred. Hα is in the red at 656.3 nm. These are the spectral features that most ground-based work on stellar chromospheres is performed with.

It is now time to introduce the concept of "local thermodynamic equilibrium" or LTE. It requires that the concept of temperature is both well-defined and local, in the context that the populations of various ionization states of elements and energy levels within atoms and molecules are what would be predicted from Maxwell–Boltzmann

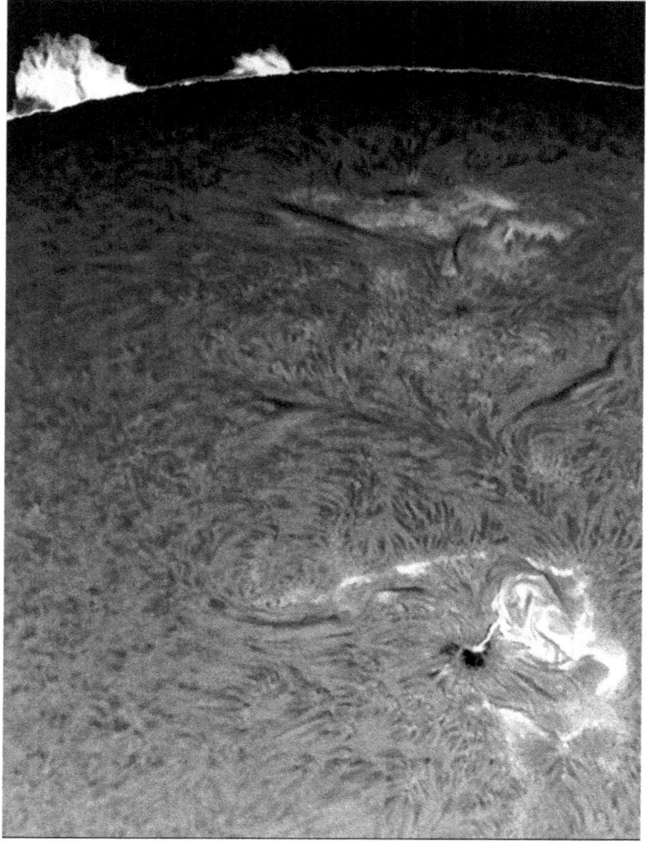

Figure 3.1. An image of the solar chromosphere taken with an Hα filter. Filaments are visible as long dark rope-like features on the disk, fibrils as the shorter dark hair-like features and prominences are brightly suspended over the limb. Active regions are visible as brighter areas on the disk. A C-class flare is taking place to the right of the sunspot at the lower right. Credit: Reproduced with permission from Greg Piepol (http://sungazer.net/bws-gallery/helios-interactive/).

statistics at that temperature. This wasn't relevant for the discussion in the previous section but was implicitly true because the densities in photospheres are generally high enough that electron collisions can enforce LTE. This can hold even if the radiation field does not strictly look like the local Planck function. It won't hold where the surrounds are not opaque enough to hide other temperatures or the presence of a surface at some wavelengths unless collisions completely dominate radiative transitions for the levels of interest.

Appendix A explains why this becomes an issue in the chromosphere and above, where the densities are no longer high enough and the spectral transitions of interest can be subject to violations of LTE. The concepts of optical depth and source function are covered there: optical depth is a way of expressing how deeply into a medium one can see and the source function expresses what intensity one sees there.

Both of these quantities depend on the level populations of the relevant atoms or molecules providing the opacity and emission or absorption. When those populations do not obey LTE the condition is called non-LTE. NLTE conditions can arise even when a particular transition might seem configured for LTE because the level populations are all tied to each other through the process of statistical equilibrium. If the reader is not sufficiently interested in these physical details to spend the considerable effort to grasp them, it is possible to gain the main points of this chapter without reading the appendix. The important principle to know is that the intensity one observes at a given wavelength is approximately the source function at optical depth unity (the depth at which a relevant photon has a good chance of escaping the star).

The manifestation of active regions in the chromosphere that is somewhat analogous to the faculae in the photosphere is called "plage." Although they are both caused by collections of magnetic flux tubes, their physical causes are somewhat different and plage can be seen over the whole disk not just near the limb. The plage is quite clear as the brighter component in a Ca II image (Figure 3.2). The spatially averaged field in plage is only of order 100 Gauss, but it is composed of individual flux tubes with fields well over 1 kG and has a filling factor well under unity in the photosphere.

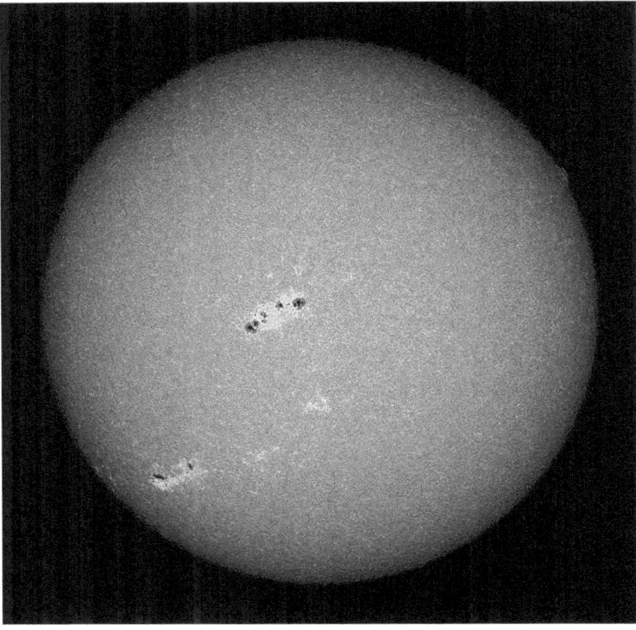

Figure 3.2. An image of the solar chromosphere taken with a Ca II K-line filter. Along with the sunspots and brighter plage, the "network" is also visible as bright outlines of supergranule cells over most of the disk. Credit: Reproduced with permission from Stephen Rahn (https://flickr.com/photos/97839409@N00/32967892013), Public Domain Dedication (CC0 1.0).

Because the magnetic pressure is increasingly dominant over the gas pressure with height, the flux tubes spread and take up more and more of the available volume, so the area covered by plage is larger than that covered by faculae and tends to be similar to the area covered by field in magnetograms. The height above the photosphere at which flux tubes will tend to fill the canopy ranges from 700-1500 km depending on the flux in the active region (lower for more flux). This is already apparent in the 1700 Å image in Figure 1.1 whose optical depth is from a metal continuum that forms in the temperature minimum region (the bottom of the chromosphere) at a height of around 500 km.

The complexity of the chromosphere is even more apparent in Hα images like Figure 3.1. The basic reasons for this is that this line is both optically thicker and formed in NLTE, so some structures appear dark and others bright depending on their height and geometrical form. Magnetic strands suspended over the denser chromosphere tend to look dark against the disk (they have lower source functions). Some of these structures can extend for thousands to tens of thousands of kilometers. When seen against the disk they are called "filaments," and when seen over the limb (where the background is dark) they are called "prominences." The lifetime of these structures can vary from minutes to weeks, so it is also important to obtain images on all these timescales. Also visible in Figure 3.1 is a flare; these only last for timescales best expressed in minutes. Hα is the best optical diagnostic for observing these magnetic explosions (discussed in Section 4.3).

Unfortunately, we cannot get fascinating and highly informative images like these on any star other than the Sun. We must therefore turn to spectroscopy and apply what we know about radiative transfer to learn details about magnetic activity and its manifestations on other stars.

3.2 Chromospheric Resonance Lines

As mentioned above, the intensity at a given wavelength in a spectrum is related the value of the source function in the atmosphere around optical depth unity at that wavelength. Appendix A defines and explains these terms; it will be helpful to skim at least its first three pages now. The optical depth depends on the number of atoms populated in the lower level of the transition of interest, integrated along the column from the observer to that depth. If the source function does not reflect the local temperature (LTE) it might contain information about other parts of the atmosphere. In any case the observed intensity will actually be a convolution of source functions weighted over the relevant optical depth integral (the contribution function; Equation (A.3)). Finally, although stellar atmospheres generally have a complex three-dimensional structure they are typically treated as one-dimensional by using a plane-parallel approximation to compute the spectrum. The primary spectral lines of interest in this section are the resonance lines Ca II H&K (393.37 (K) and 396.85 (H) nm), Mg II h&k (279.55 (k) and 280.27 (h) nm), and Lyα (121.57 nm). The physics behind the formation and appearance of these resonance lines (that all have the ground state as their lower level) is the subject of Appendix A.2; refer there if some of the discussion below is otherwise mysterious.

Let us now consider the formation of a line like the Ca II K resonance line. We are interested in it because we know it is quite optically thick, enough that the core will be formed (reach optical depth unity from above) fairly high in the chromosphere for solar-type stars. Part of the reason for this is the abundance of calcium, and part is due to the ionization state induced by the chromosphere (calcium becomes singly ionized). The Einstein A value for this transition (Appendix A.1) is also quite high ($\sim 10^8$)—it is a "strong" transition. That means it has strong damping wings so the solar spectrum shows the effect of them several nm away; they run into the H-line damping wings between the two lines. Of course there are many other spectral lines embedded in these damping wings, but the overall line shape is responsive to the behavior of the atmosphere in the upper photosphere starting when the wings are quite weak, the temperature minimum is responsible for the minima at the outer edges of the emission core, and inward of that to line center is formed in the chromosphere. Examples of Ca II H&K in an active and inactive star are shown in Figure 3.3.

The source function for the K-line is NLTE because of its strong scattering character (the level populations are controlled by non-local photons) and complete frequency redistribution (Appendix A) is a poor assumption, so care must be taken in inferring physical atmospheric parameters from the line intensities (cf Appendix A.2). The effect of this is to raise the actual temperature inferred from the line profile compared with what would be inferred assuming LTE. For example, the solar temperature minimum implied by the depth of the line is a few hundred degrees higher than LTE would imply (Ayres & Linsky 1976). The slope of the wings inward

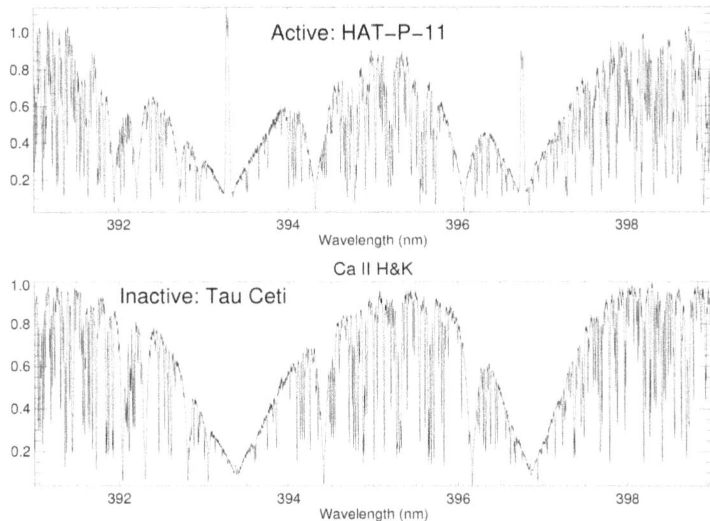

Figure 3.3. Two spectra of the Ca II H&K lines taken with the High Resolution Echelle Spectrometer (HIRES) at the Keck Observatory. The upper one is more active and shows strong emission cores. The damping wings of the calcium lines are riddled with other photospheric lines (mostly from iron). Credit: Howard Isaacson and the California Planet Survey.

in wavelength toward the emission core provides information about the gradient of decreasing temperature with increasing height in the upper photosphere. Due to the greater heat capacity at higher densities the actual temperature continues to drop until the density has dropped sufficiently even though non-radiative heating is present, then begins to rise with height above the temperature minimum. The non-radiative heating in the upper photosphere causes its temperature gradient to be shallower than radiative equilibrium would suggest. The temperature minimum is hotter in stars that have more active chromospheres than in comparable but less active stars.

A set of observed stellar K-lines with varying levels of stellar magnetic activity is shown in Figure 3.4. The full disk solar spectrum is much like that of Gleise 17 in the upper left panel. Because most of the Sun is not very magnetically active, the K-line in integrated sunlight barely registers the presence of the chromosphere. It is only indicated by the slight bump in the downward trend of the left wing before the final drop at line center. The red emission peak is almost invisible in the upper panel row (the symmetry of the core can be influenced by flows in the atmosphere). The subtlety of these features is unfortunate because it makes studying magnetic activity on stars like the Sun (or even less active stars) rather difficult in this chromospheric diagnostic (and also in Hα for different reasons). The presence of emission near line center is a little more obvious in the second row down of more active stars, and quite obvious below that.

Notice that there is a dip in the very center of all the lines (inward of the emission). The line opacity increases all the way to line center, but this reversal does not mean that the temperature drops again above a certain height. The fact that the emission is not as bright as the wings also does not mean that the chromosphere is not as hot as the inner photosphere. Rather the central dip is due to the fact that the NLTE source function has departed from the local temperatures where the line opacity is thick enough to probe these lower densities. The kinetic temperature associated with the atmospheric region where this central "absorption" core is seen is actually still increasing outward and is higher than the temperature where the highest line emission originates. The formation of this spectral resonance line is conceptually much like the situation in the bottom middle panel of Figure A.1 in the Appendix. In active stars cooler than about 3500 K the neutral potassium resonance doublet at 766.5, 769.9 nm shows similar behavior and line profiles.

Metrics of chromospheric strength are often defined by the excess emission over a star with no chromosphere (a radiative equilibrium photosphere). Alternatively one might define the excess over a basal level of chromospheric heating due to purely acoustic heating and not magnetic in origin. From an observational point of view, the task is to obtain the spectrum of a comparison star with the same effective temperature and chemical composition and subtract it from the spectrum of interest. The wings of the calcium lines are much less sensitive to changes in magnetic field than the core, so the task can be reduced to finding a star whose wings match and looking at the residual core emission. If the stars in Figure 3.4 were all sufficiently similar, one could use a spectrum in the upper row as the template to be subtracted (adjusting for radial velocity differences). It is clear that the core emission in all the

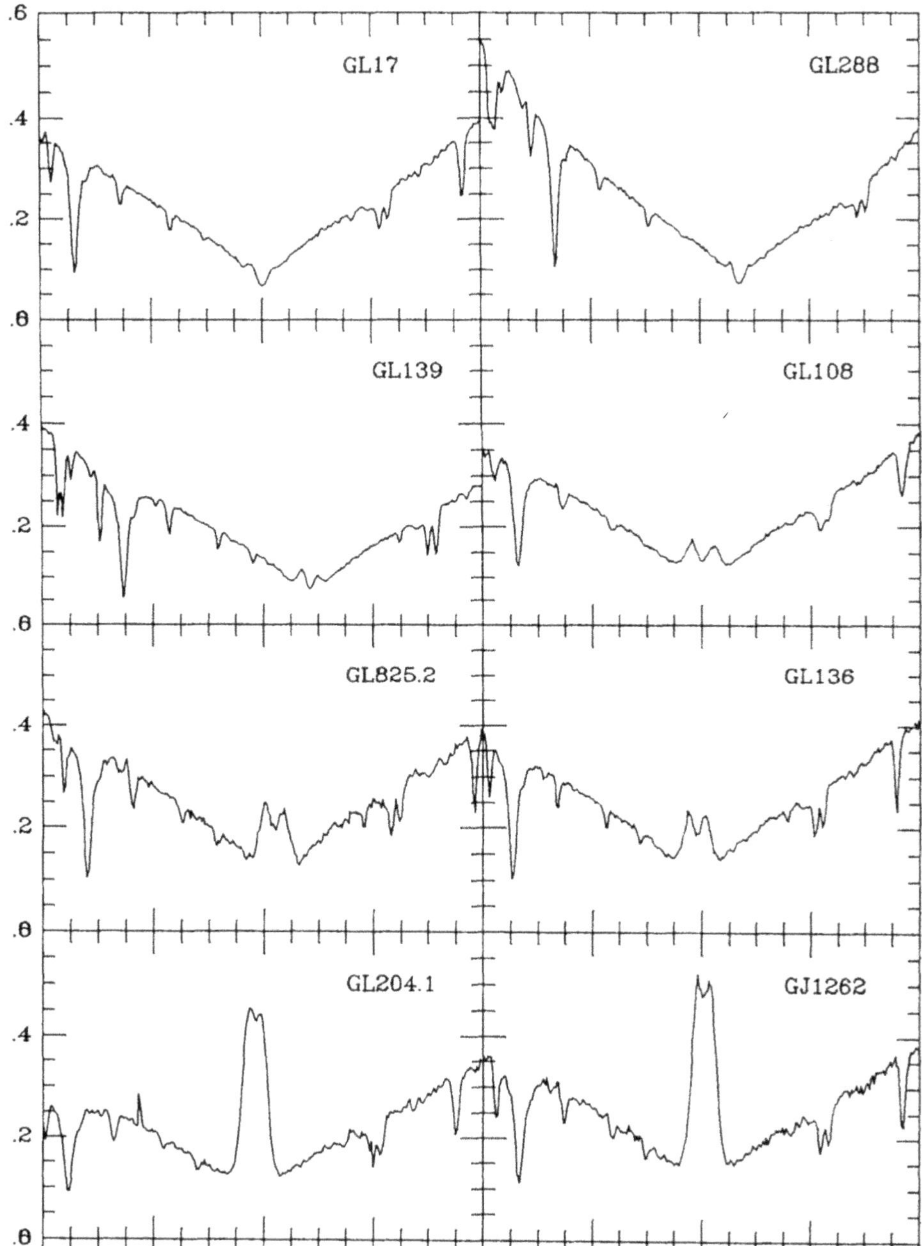

Figure 3.4. A set of observed stellar K-lines cores exhibiting varying levels of magnetic activity. The top two are like the Sun (which barely manages to show any core emission in integrated light). Moving downward the stars are increasingly active; they resemble what is seen on increasingly active areas on the Sun. The vertical scale is relative to the maximum flux between the H- and K-lines, and the horizontal major ticks are each 0.2nm, centered at 393.3nm. Credit: Pasquini (1992), reproduced with permission © ESO.

others would then show up as a residual excess; that excess becomes larger in lower rows and the subtraction of the template fractionally less significant. An alternative is to use spectra from model atmospheres without non-radiative heating as the templates (these would not contain even a basal flux).

An interesting effect was noticed about the core emission of the H&K lines even in objective prism spectra early on. The width of the base of the core emission feature (or its FWHM) becomes wider as the absolute luminosity increases or equivalently as the surface gravity decreases in a diverse sample of stars with calcium emission (Figure 3.5). This is known as the "Wilson–Bappu" effect for its discoverers in 1957, and explaining it was one of the early motivations for studying stellar activity. The modern explanation for this effect was developed by Ayres (1979). The location of the emission base can be associated with the atmospheric temperature minimum. When computing the K-line using proper NLTE radiative transfer with partial redistribution, the gravity dependence of the wavelength of the base of the emission can be related to the mass column density of the chromosphere. Ayres (see also his chapter 2 in Engvold et al. 2019) makes detailed arguments about the balance of radiative cooling (in both strong lines and continua) and the non-radiative heating that generates the chromosphere. These relations are density-dependent, and translate to a gravity dependence because lower gravity stars will have lower densities and thicker chromospheres. The radiative transfer in the K-line then translates that to the luminosity-width effect in the spectrum.

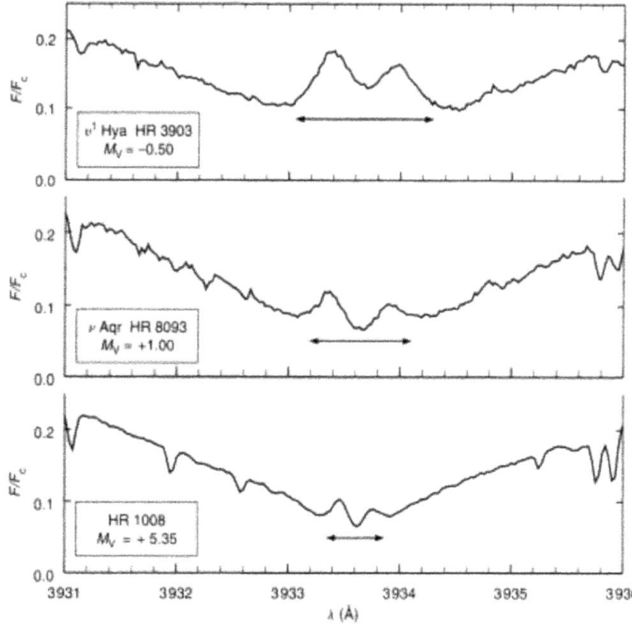

Figure 3.5. The Ca II K line observed on 3 stars with of very different absolute luminosity (and so surface gravity). The amount and asymmetry of the emission features are similar, but the widths vary with luminosity. This variation is known as the "Wilson–Bappu" effect. Credit: G. Pace, ESO.

A much better pair of lines for studying chromospheric activity are the Mg II h&k lines, which unfortunately are in the vacuum UV and so can only be seen from space. Figure 3.6 show a representative set of profiles from quiet and active dwarfs, and a pair of more active RS CVn systems (somewhat evolved close binaries in synchronous orbit). It is immediately clear from the solar analog (and closest neighbor) α Cen A that despite the fact it is even a little less active than the Sun, the Mg II h&k lines exhibit strong emission that is very easy to measure. This is partly due to the fact that these lines are optically thicker than the calcium lines (the abundance of Mg is about 15 times that of Ca) and so their cores are formed higher in the chromosphere, but mostly because the photosphere is relatively fainter and the chromosphere relatively brighter because of the behavior of the Planck function given the shorter wavelength and the temperature contrast. The central NLTE reversal is absent on the more active cases.

Figure 3.7 shows a set of Mg II profiles measured on more evolved stars. The central NLTE reappears quite consistently when observing stars of lower surface

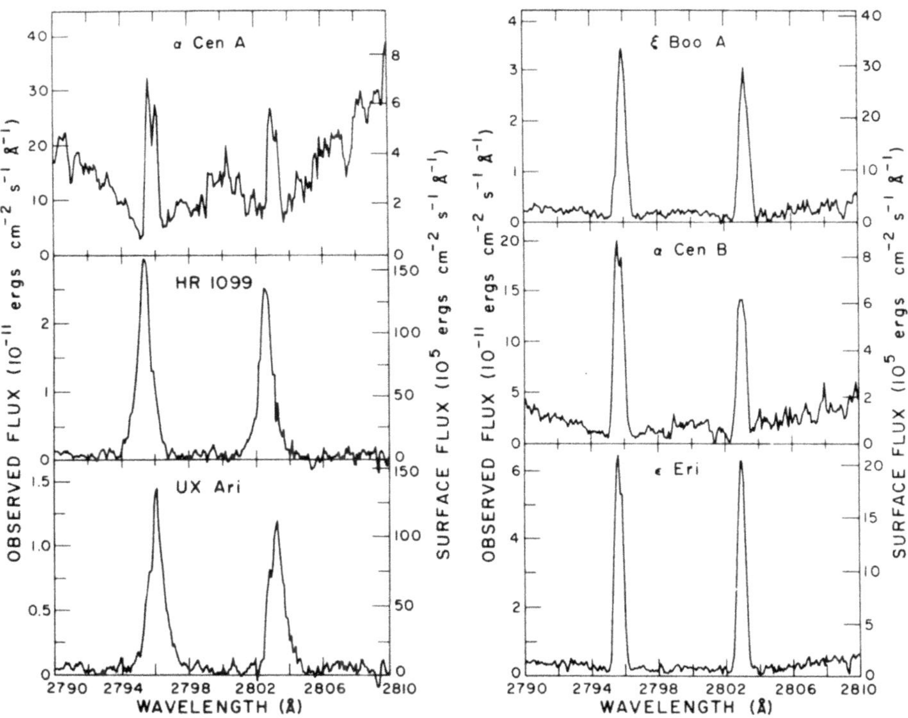

Figure 3.6. A set of Mg II h&k lines observed using the International Ultraviolet Explorer (IUE) spacecraft with varying levels of magnetic activity on the stars. In the upper left, α Cen A is much like the Sun, and α Cen B has the same age and composition, but is a little less massive. ξ Boo A and ε Eri are more active main sequence stars. UX Ari and HR 1099 are RS CVn systems; they have more evolved components in close binaries. Credit: Reproduced from Basri & Linsky (1979). © 1979. The American Astronomical Society. All rights reserved.

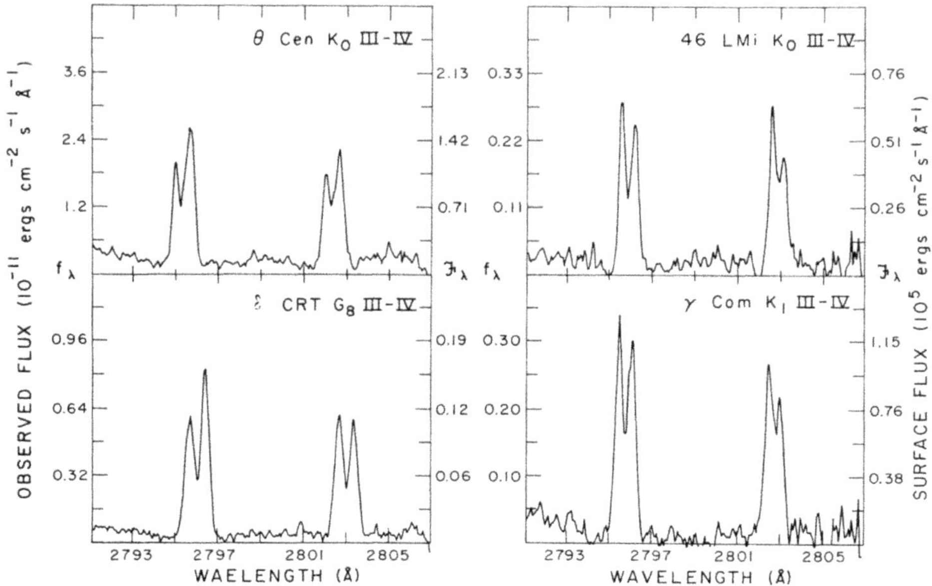

Figure 3.7. A set of Mg II h&k lines observed using the IUE spacecraft on stars that have left the main sequence. The central reversals are now all obvious. Asymmetries between the emission peaks are diagnostic of bulk flows at the top of the chromosphere. Credit: Reproduced from Stencel et al. (1980). © 1980. The American Astronomical Society. All rights reserved.

gravity because their densities are lower and NLTE is more prevalent. The analog of the Wilson–Bappu effect is can be seen when comparing this figure with the last; the main sequence stars have narrower emission profiles. The asymmetries in the red and blue peaks are good diagnostics of upflows (red peak higher due to absorption in the blue peak) or downflows (opposite). Many studies (cf Linsky 2017) have shown that the emission excesses in the calcium and magnesium lines are closely correlated with each other (and with other diagnostics of stellar activity). It is a pity that we have not yet had a mission dedicated to measuring the Mg II emission from a large number and variety of stars; it would be even more useful if it provided information on temporal variability at short and long timescales. That would be one of the cleanest ways to gather trustworthy data on the behavior of stellar activity over mass, age, and rotation.

The formation of the ionized magnesium resonance lines is similar to that of the ionized calcium resonance lines. Figure 3.8 shows the appearance of the Sun in Mg II k. The image in the wing (less optically thick) shows many small bright points in the upper photosphere that reflect local sites of strong heating. The spreading of the heated plasma is obvious in the core of the line which shows much greater bright areas in the upper chromosphere. These are accompanied by spectra of the line at the indicated colored locations: plage profiles on the left and profiles outside plage on the right. The observations show that although the average profile (dashed line) shows a small central NLTE reversal, it can disappear when the density is high

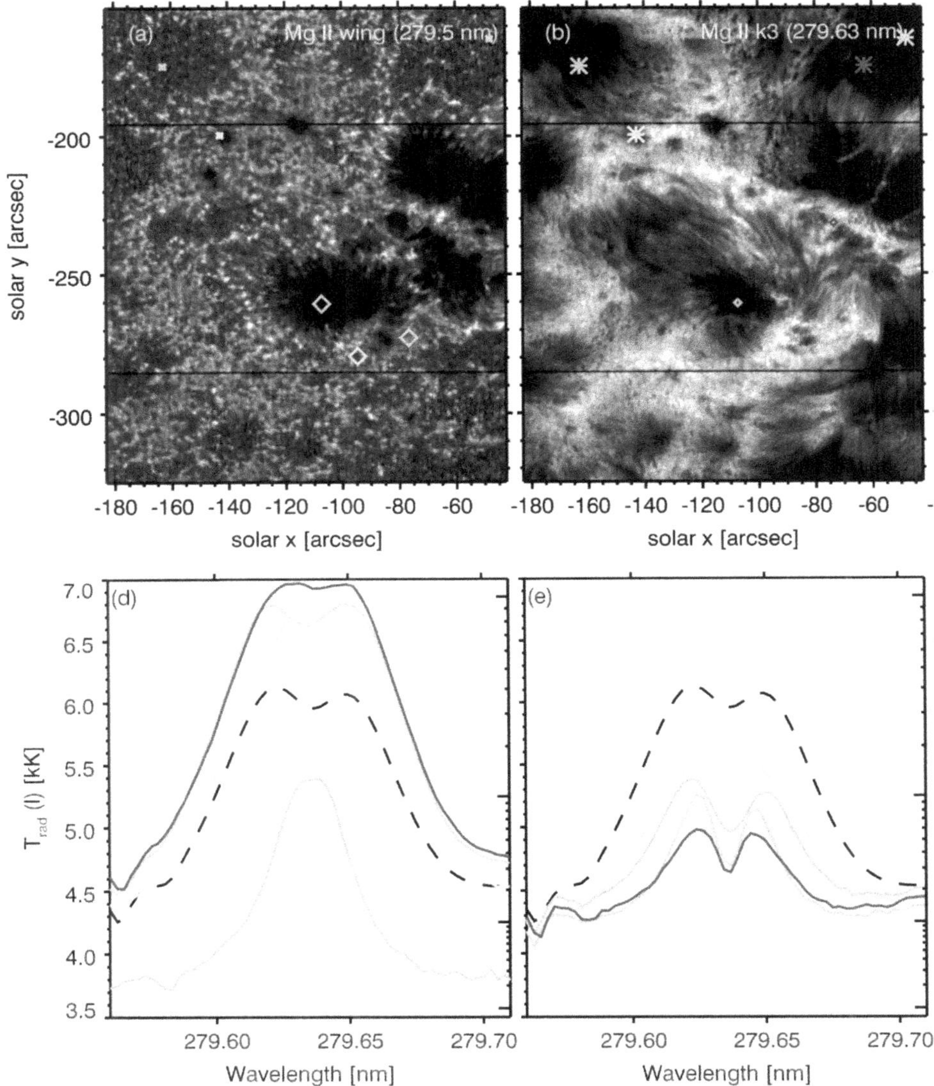

Figure 3.8. The Mg II k line observed on the Sun using the Interface Region Imaging Spectrograph (IRIS) spacecraft. The left image shows the upper photosphere in the wing of the line (less optically thick) and the right image the upper chromosphere in the core of the line. The colored spectra are taken at the location of the colored symbols, and the average profile over the whole image is the dashed line. Credit: Reproduced from Carlsson et al. (2015). © 2015. The American Astronomical Society. All rights reserved.

enough in the upper chromosphere (the source function is closer to LTE). Modeling of the solar lines has progressed to the point where 3-D effects can be combined with sophisticated radiative transfer models (Carlsson et al. 2015).

Lastly we discuss the resonance line of neutral hydrogen: Lyα. Since hydrogen is by far the dominant constituent of stars. For solar-type or cooler stars it will largely be neutral until becoming partially ionized in the upper chromosphere, and mostly in the ground state because of the large energy jump to level 2. Thus Lyα is the most opaque spectral line in the stellar atmosphere. It therefore gives us the highest altitude view of the stellar atmosphere so long as hydrogen is neutral (it is hardest to see into the star at this wavelength). However, because the transition region above the chromosphere has an extremely steep upward temperature gradient that ionizes hydrogen, the core of Lyα is formed not much higher up (a few hundred km) than the Mg II lines. Indeed, the temperature gradient is so steep because of the loss of radiative cooling through hydrogen channels (both bound-bound and bound-free). Recall that radiative cooling occurs when a transition is excited by an electron collision, thus taking kinetic energy out of the thermal pool, and that energy is radiated away if the transition de-excites by emitting a photon that is lost from the region. This only works if such transitions are available in the necessary quantity to overcome the heating that is present.

Because of its very high Einstein A coefficient and very high ground state population, the damping wings of Lyα are also quite opaque so the wings are extended. The main point below is that very strong lines must be treated quite carefully to correctly infer atmospheric properties from their observed profiles. The rest of this paragraph will be obscure without knowing what is in Appendix A but is not crucial to what follows. Its atomic properties mean that this line has a very small ε and so partial frequency redistribution is essential to interpreting the wing intensities and shape (cf Appendix A.2). Although coherent scattering smeared by a Doppler width is a fair description of wing source functions, it turns out that the small contribution by partial redistribution of the line core source function does play a role. The line core is thermalized at depths where the wings are not, so some local thermal information leaks in. It is also the case that these lines provide significantly smaller radiative cooling to counteract non-radiative heating than would be estimated by assuming complete redistribution. Basri et al. (1979) discuss the formation of the solar Lyα line profile in detail.

When observed with high spatial resolution the Sun displays quite a range of profiles as shown in Figure 3.9 (this is true for all chromospheric lines). The "solar analogy" that is often used in trying to understand other stars is an assumption that more active stars are simply covered by larger fractions of increasingly active areas like those seen on the Sun. This does seem to be true at a useful level, although we should always keep in mind that it is unlikely to be true in detail. Because we don't actually have images of the surfaces of other stars, it is hard to know how the averaging that takes place of myriad atmospheric structures on a given star translate into its observed integrated flux. Even on the Sun, the reproduction of spatially resolved profiles by semi-empirical model atmospheres is hampered by the assumption of a plane-parallel atmosphere (Section 3.5). It is not surprising, therefore, that models constructed from different diagnostics can differ in sometimes significant ways.

For other stars, a major problem with Lyα is the presence of interstellar absorption by neutral hydrogen (Figure 3.10). This eats away a major chunk near line center, whose width depends on the column density and the velocities of the various possible H I clouds between us and the star. It can therefore be difficult to

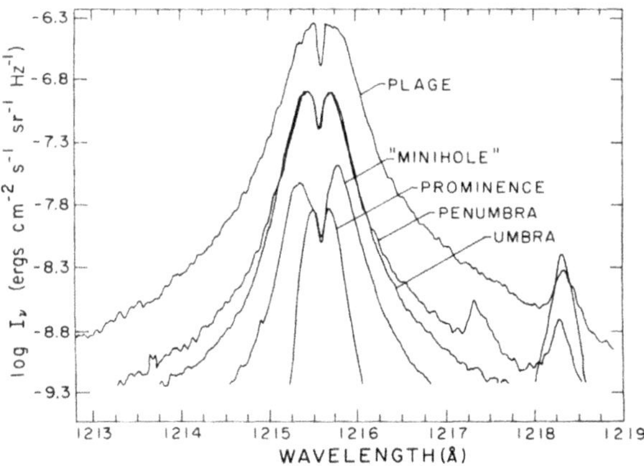

Figure 3.9. Observations of the Lyα at various points on the Sun from a rocket experiment. Credit: Reproduced from Basri et al. (1979). © 1979. The American Astronomical Society. All rights reserved.

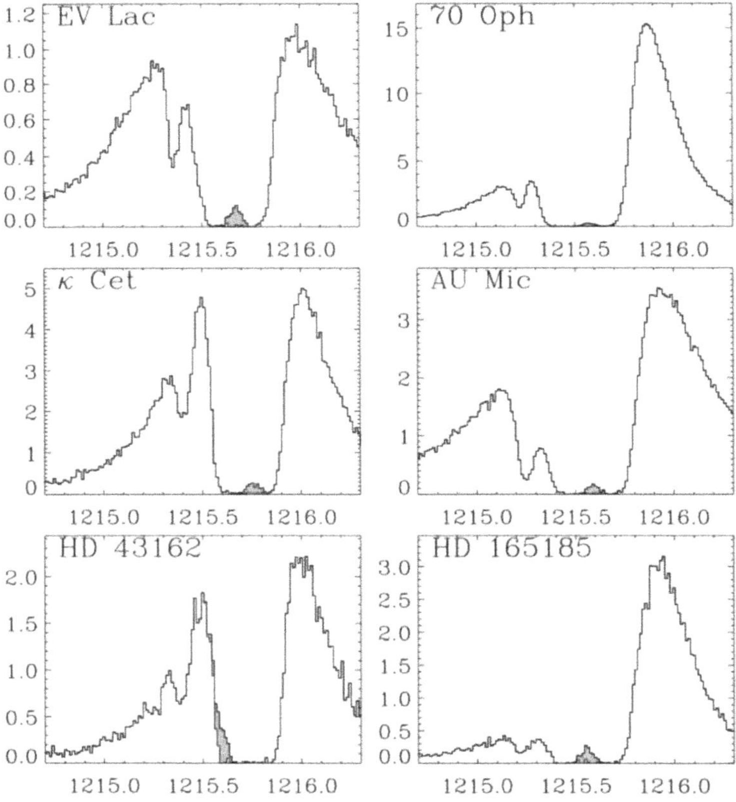

Figure 3.10. Observations of Lyα collected by the Hubble Space Telescope (HST) on a number of nearby stars. The large flat-bottomed central absorption is due to the local ISM; the smaller absorption about 0.33Å blueward is from interstellar deuterium. The grayed emission is from the geocorona. The vertical scale is in observed flux units and the horizontal scale is Å. Credit: Reproduced from Wood et al. (2005). © 2005. The American Astronomical Society. All rights reserved.

know what the total Lyα flux from the star is; this has become of particular interest because of its likely major effects on exoplanet atmospheres (especially around red dwarfs and young stars). One recent discussion of this topic is in Linsky et al. (2020); who also refer to the extensive previous work on utilizing Lyα to measure the local interstellar medium (ISM; Mg II is also useful for this) and the interstellar deuterium abundance. All these analyses require making some estimate of what the underlying stellar chromospheric line profile would look like without the ISM absorption. Making matters slightly worse, the Earth is surrounded by a cloud of neutral hydrogen activated by the Sun that produces geocoronal Lyα emission. A discussion of the corrections needed can be found in Wood et al. (2005).

Finally the He I (58.4 nm) and He II (30.4 nm) resonance lines are also important in principle, but are formed in a part of the spectrum that is very difficult to observe even on the Sun. They are largely inaccessible to us from other stars due to the opacity of the ISM in the Lyman continuum, although a few observations were made by the Extreme-Ultraviolet Explorer (EUVE). Furthermore the formation of helium lines is influenced by both high-energy radiation and collisional excitation and ionization by high-energy electrons from the transition region and corona. These are more energetic than expected at the local region of line formation since both the radiation and particle distributions have a mix of temperatures. Both the chromosphere and the corona play a role even for the observationally accessible non-resonance lines He I 587.6 nm and 1083.0 nm, making their interpretation more difficult (and still somewhat controversial). They do show general correlations with stellar activity and so are somewhat useful in that respect. A more detailed discussion of the He I infrared line is at the end of the next section.

3.3 Hα and Other Diagnostics

The line that is most associated with studies of magnetic activity on both the Sun and stars is Hα. As mentioned before this is the level 2–3 transition in neutral hydrogen at 656.28 nm. It was identified by Fraunhofer in the early 19th century, and George Ellery Hale invented the spectroheliograph to make images in the line by the end of that century. As photography of the Sun came into its own the Hα line was found to be the best means of imaging the chromosphere. It is also the source of the red color that gave the chromosphere its moniker. As stellar spectra began to be gathered it was also clear that some stars exhibit Hα emission. The designation "e" was added to spectral types to indicate its presence, although that is not always an indication of magnetic activity. The first dMe flare stars, V1396 Cyg and AT Mic were formally designated in 1924.

Despite its popularity as a chromospheric diagnostic on the Sun and red dwarfs, the spectral line formation physics of Hα is actually quite complicated and less easy to interpret than most spectral lines. It is not a resonance line and its lower level has an excitation energy significantly above the ground state (relatively speaking). The temperature of the solar photosphere is inadequate to collisionally excite much of a population in level 2; that requires either excitation by the Lyα line in the far-UV, or a cascade down from higher levels after a recombination of a proton and electron.

The radiation needed to previously ionize hydrogen before recombination is the Lyman continuum in the EUV because the ground state is heavily populated. Appendix A.1 explains that the source function can be written as a ratio of level populations. It is clear that the level 2 and 3 populations of hydrogen will have several complicated processes controlling their ratio. This means that Hα is very much formed in NLTE (Appendix A.2) and so is not very indicative of the plasma temperatures at optical depth unity at various wavelengths in the line. This is the reason that Hα is a pure absorption line in solar spectra for all but the most magnetically active locations despite its core always being formed in the chromosphere.

Despite the fact that the Hα source function is controlled more by radiation fields than the local Planck function, it still looks different in various active components on the Sun. These differences arise both from the decoupling (from LTE) parameters induced by the different densities and temperatures in these structures, and also the changes in optical depth scales. The same optical depth at a given wavelength might probe above the temperature minimum in one structure and below it in another. Figure 3.11 shows a case on the Sun. Notice that in an image taken about 0.5 Å off line center, a plage could actually appear darker than the dark points (this is observed). Further work would have to have been undertaken on the particular models shown (Basri et al. 1979) to reconcile their mis-match in the outer wings if that had been the goal of the paper. In fact those models based on Lyα were later shown to suffer from issues related to the formation of silicon continua. It illustrates how complicated an analysis of NLTE lines might have to get, how entangled with

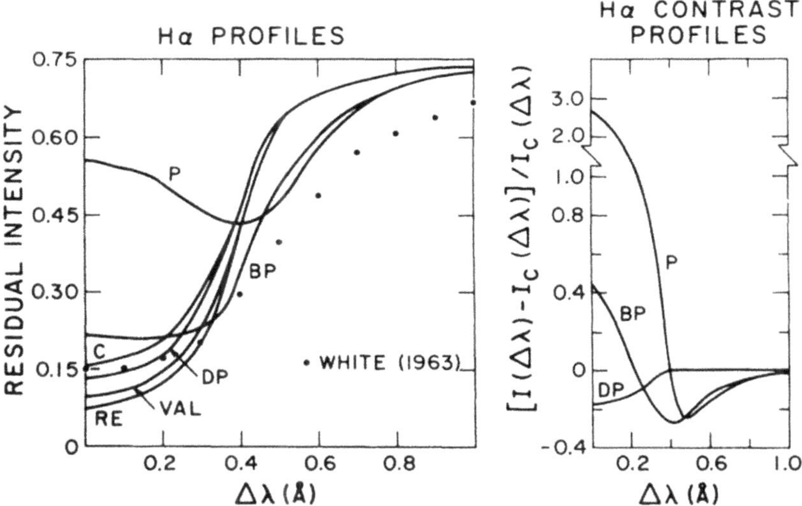

Figure 3.11. The left panel shows set of computed Hα static half-profiles from various models; observations are shown with the dots. The standard solar model is VAL, a radiative equilibrium model is RE, a plage model is P, dark points are DP and bright points are BP. On the right the wavelength-dependent contrasts between each profile and that for the average model C are shown. Credit: Reproduced from Basri et al. (1979). © 1979. The American Astronomical Society. All rights reserved.

many transitions and species the full modeling problem can get, and how difficult it can be to really understand what disk-integrated line profiles really represent. In fact models can never be fully reconcilable with observations when utilizing 1-D plane-parallel atmospheres (Section 3.5).

The same complications apply to the interpretation of Hα in stellar spectra along with a set of further issues. The level of magnetic activity that produces an emission line is a function of stellar effective temperature. This is partially due to the fact that hotter stars will have an easier time populating level 2 of hydrogen (the lower level for Hα), and partially due to the fact that whether the core appears in emission or absorption is a contrast issue. A certain flux in the line core will appear in absorption (be darker) against a continuum flux that is brighter, but in emission compared with a continuum flux that is fainter. These two effects explain why the line is in absorption for all but the most active F, G, K stars. The depth of the absorption is therefore complicated to interpret, which means that Hα is not a great activity diagnostic for solar-type stars.

When considering spectral type M one would not expect to see an Hα line at all due to nearly complete depopulation of level 2 at those cooler photospheric temperatures. The presence of a chromosphere, however, means that level 2 will be populated in that part of the atmosphere. Chromospheres lie between 6000-10000 K in order that there be sufficient radiative cooling to balance the magnetic heating; this is where efficient cooling transitions operate. These effects lead to a counter-intuitive behavior of the observed Hα spectral line with effective temperature for late K and M stars. For a given spectral type the line is weakly in absorption if there is little magnetic activity, grows deeper in absorption with increasing activity because level 2 is more populated, becomes weaker again at higher levels of activity as the core brightens because chromospheric densities are increasing (coupling it more effectively to the chromospheric temperatures), and eventually goes into emission at high enough activity levels.

This scenario plays out differently depending on the stellar effective temperature (and surface gravity). It is desirable to have an independent measure of the activity level to confirm it. That can be provided by Ca II or X-rays or some other measure that is less ambiguous than Hα. It is also helpful to make the observations in different diagnostics at the same time since activity levels can vary over time. Walkowicz & Hawley (2009) provide one demonstration of the analysis of Hα and other diagnostics, and reference previous work including the original modeling. More recent efforts using alternate indicators of activity include those of Schöfer et al. (2019), Hintz et al. (2019), and Tilipman et al. (2021).

Another instance where Hα provides a helpful diagnostic of plasma with chromospheric temperatures is when it is seen in the same way as filaments are seen on the Sun, namely in absorption due to prominences against the disk. When M stars experience large flares Hα will go in emission, but one can also sometimes see a temporary blueshifted absorption component superposed on that due to cooling plasma being ejected. Rapidly rotating active stars like AB Dor also show features interpreted as prominences being carried around by stellar rotation. Jardine & Collier Cameron (2019) discuss such observations and also provide more detailed modeling of how these features are formed, adopting the terminology "slingshot prominence." If a prominence forms above the sonic point and at the Keplerian co-rotation radius (where the effective gravity is close to zero) then upflows through the loop will become supersonic before reaching the top. As the density grows there the loop will have more trouble containing the plasma, and once it grows past the Alfvén radius it will pull open and spill its contents outward. This topic is pursued in Section 4.4.

The Ca II IR triplet lines near 850 nm are a well-utilized example of other lines that are in absorption until the activity is very strong and pushes them into emission. Their formation physics is different from Hα (less NLTE) and they sample lower in the atmosphere, but they can be used as a metric of activity in appropriate circumstances. They resemble Hα in the sense that they are generally absorption lines, but fill in as activity levels increase and are in emission in the most active cases. One method of extracting activity information is to subtract the line profile from an inactive star with otherwise similar parameters. An adjustment for the rotational broadening of the more active stars might be needed since activity increases with rotation for stars of similar spectral type. An early example of this technique is in Linsky et al. (1979). An exploration of other strong lines formed in the upper photosphere and lower chromosphere shows that there are a number of lines that can be used to probe activity in a similar way (Sasso et al. 2017). Basri et al. (1989) conducted a broad search in echelle spectra for other photospheric lines that show residual emission in subtracted lines and found a few. A recent study of α Cen B at various stages of activity (Thompson et al. 2017) at higher resolution and signal-to-noise identified even more of them.

The effect of magnetic activity on photospheric spectral lines has garnered more recent interest in the context of precision radial velocity searches for exoplanets. There are three main effects of stellar activity that can affect a precision radial velocity measurement. One is the presence of starspots that can cause parts of the stellar disk at particular velocities to be fainter and thus affect a line flux profile. Another is that lines can be partially filled in by upper photospheric heating as discussed above which changes their weighting in an auto- or cross-correlation analysis, and their velocity profiles could be different than average. Finally, all lines are affected by the "convective blueshift" that arises because the hotter brighter interiors of granules are rising while the dark lanes contribute less light but move more rapidly downward. The presence of magnetic activity affects these flows and light balance and so the line asymmetries. There is a large and increasing literature on this topic, for example Meunier et al. (2017), Wise et al. (2018), and Dumusque (2018).

Another line whose formation is in both the chromosphere and transition region and whose formation mechanisms are both complicated and NLTE is the near-infrared He I 1083.02 line. The lower level of this transition is metastable, and can be populated either by recombination cascade from He II or by collisional excitation of lower levels of He I. Of course it takes a lot of energy to ionize helium, so the driver of the recombination population is actually EUV or X-ray photons from the upper transition region or corona. Because the level is metastable it will also have a higher population at lower densities (other factors being the same). This line usually appears in absorption but a stronger line means larger high-energy fluxes. The collisional channel requires relatively high temperatures to raise He I to that level of excitation so it is more diagnostic of conditions in the middle transition region and will become stronger if densities are greater. These complications mean that a careful and sophisticated analysis is needed to extract the most accurate physical information from this line. On the helpful side it is diagnostic of very hot plasma without requiring space observations of short wavelength light. It also turns out to be useful in diagnosing hot plasma from accretion and winds in T Tauri stars (Section 7.2), and has also been seen weakly in absorption in metal-poor giants that are not likely to have strong magnetic fields. A more detailed discussion is given by Linsky (2017).

3.4 Semi-empirical Chromospheric Models

Linsky (2017; Section 5) provides an extensive discussion and listing of the history of assumptions, methods, and results in the literature on semi-empirical chromospheric models. Almost all models use 1-D plane-parallel geometry and assume statistical and hydrostatic equilibrium. Another good description is provided in chapters 3, 5 of Engvold et al. (2019), which also has some information about 3-D models. Some solar models have extensive input and modeling from a variety of continuum and line diagnostics and separately model different regions on the Sun with different levels of magnetic field. The standard for a long time were the VAL models from Vernazza et al. (1981); the canonical version is shown in Figure 3.12. These incorporate NLTE physics and partial redistribution in spectral lines as needed (Appendix A), but also rely on the center-to-limb behavior in a variety of continua. As a segue to the next chapter I note that the transition region starts so abruptly and is so thin that statistical, thermal, and kinematic equilibrium are no longer fully valid assumptions in that part of the atmosphere. Particle mean free paths can traverse significant parts of the temperature gradient, depositing particles of the "wrong" kinetic energy into the local plasma. This creates issues with the computation of level populations, ionization states, and collisional rates. The real plasma thus does not conform to physical states inferred from a 1-D atmosphere.

A particular example of this is presented by Fontenla et al. (1991). They take account of non-Maxwellian particle distributions, and ambipolar diffusion to account for the fact that neutral atoms (particularly hydrogen) can cross magnetic field lines while ions can't, but they collide with each other. This means that there is significantly more neutral hydrogen near the base of the transition zone than equilibrium calculations suggest, and that resolved problems with semi-empirical models that required temperature plateaus in that part of the atmosphere that would be very hard to maintain physically. They repeat the exercise of VAL and produced a grid of models of the solar chromosphere for different activity levels (Figure 3.13). Notice that the changes that produce brighter emission in chromospheric diagnostics involve heating of the upper photosphere, raising of the temperature plateau in the chromosphere, and moving the base of both the chromosphere and transition region

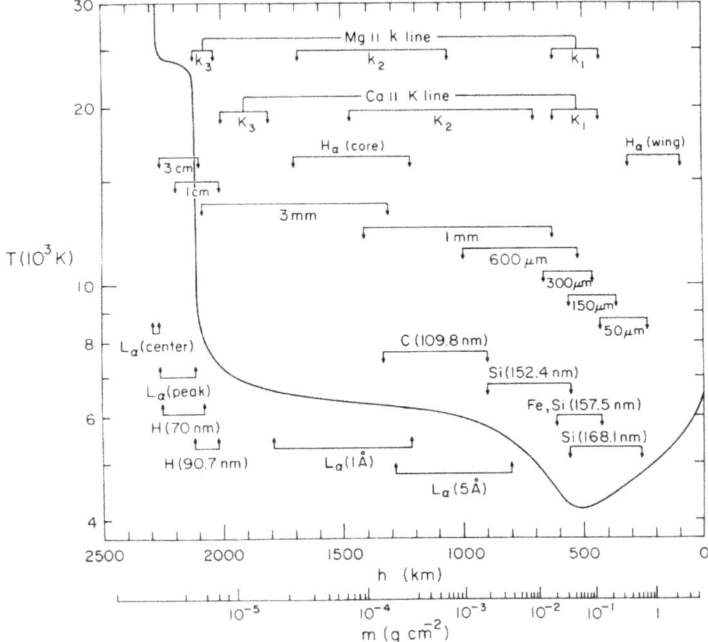

Figure 3.12. The standard semi-empirical solar atmosphere VALIII from Vernazza et al. (1981). The diagnostics used and the regions of the atmosphere where their contribution functions lie are indicated. The height scale is measured from continuum (Rosseland) optical depth unity. The paper also presents models like this for different activity levels on the Sun. Credit: Reproduced from Vernazza et al. (1981). © 1981. The American Astronomical Society. All rights reserved

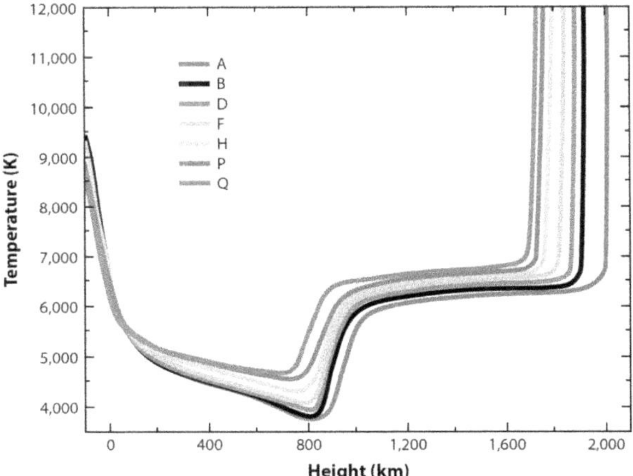

Figure 3.13. Extensions of the VAL semi-empirical solar atmospheric models that drop some equilibrium assumptions (note the height scale is reversed from the previous figure). They are produced for regions with different levels of magnetic activity (increasing from A to Q). Credit: Reprinted from Linsky et al. (2017), © Annual Reviews, who adapted it from Fontenla et al. (1991).

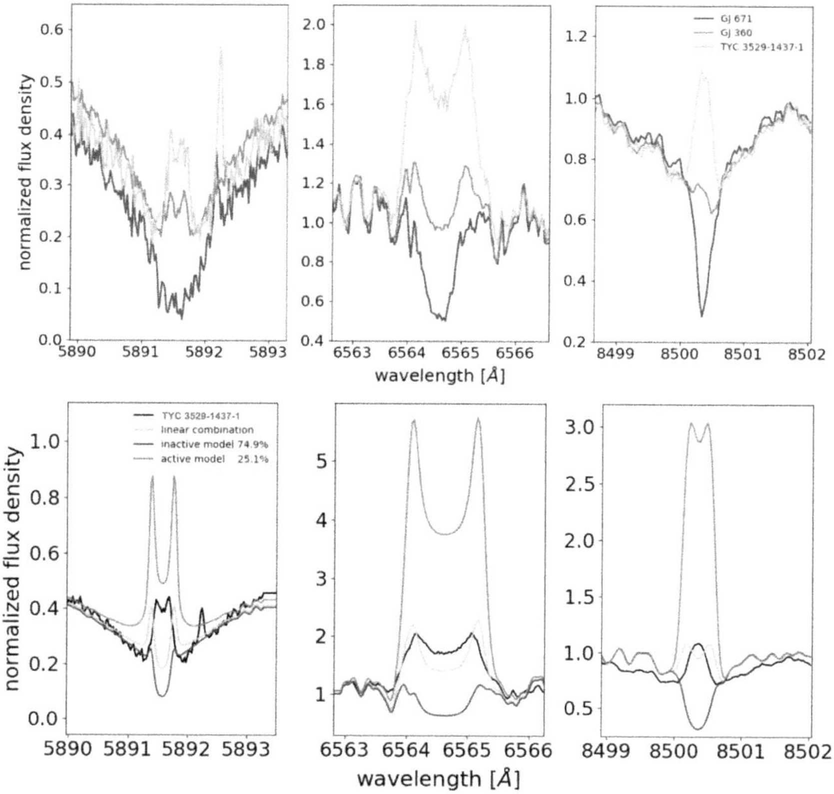

Figure 3.14. The upper panels shows a set of simultaneous line profiles from three M dwarfs with effective temperatures around 3500 K and different levels of activity. The left panels have one of the Na I resonance lines, the middle panels show Hα, and the right panels show one of the Ca II IR triplet lines (using vacuum wavelengths). All three lines are in absorption in the most inactive star; the other two stars are fairly active. The lower panels show attempts to reproduce the observations with profiles by mixing models with different activity levels. Credit: Hintz et al. (2019), reproduced with permission © ESO.

to higher mass column densities (lower heights). These tend to be qualitative characteristics of similar model sets for other stars although those have different temperatures and height and column density scales. It is worth recalling again that the stellar surfaces almost certainly are highly complex (Figure 3.8) and these 1-D models are only indicative of certain general behaviors. Truthfully we really don't know at all how stars much more active than the Sun would look if we had solar spatial resolutions.

Examples of papers that study the behavior of diagnostic lines in non-solar-type stars utilizing semi-empirical models are Walkowicz & Hawley (2009) and Hintz et al. (2019). Observations and models of some of the spectral lines discussed in Section 3.3 are shown in Figure 3.14. The models provide a quantitative estimate of at what mass column density the chromosphere must penetrate in order to produce lines with varying amounts of emission, and an estimate of the chromospheric temperature gradient. These models do not raise temperatures in the upper photosphere or have different gradients in the chromosphere, while others do. An extensive list of attempts through 2015 to model spectral lines to understand and compare the physical conditions in stellar chromospheres is provided by table 4 in Linsky (2017).

References

Ayres, T. R. 1979, ApJ, 228, 509

Ayres, T. R., & Linsky, J. L. 1976, ApJ, 205, 874

Basri, G. S., & Linsky, J. L. 1979, ApJ, 234, 1023

Basri, G. S., Linsky, J. L., Bartoe, J. D. F., Brueckner, G., & van Hoosier, M. E. 1979, ApJ, 230, 924

Basri, G., Wilcots, E., & Stout, N. 1989, PASP, 101, 528

Carlsson, M., Leenaarts, J., & De Pontieu, B. 2015, ApJ, 809, L30

Dumusque, X. 2018, A&A, 620, A47

Engvold, O., Vial, J.-C., & Skumanich, A. 2019, The Sun as a Guide to Stellar Physics (Berlin: Springer)

Fontenla, J. M., Avrett, E. H., & Loeser, R. 1991, ApJ, 377, 712

Hintz, D., Fuhrmeister, B., Czesla, S., et al. 2019, A&A, 623, A136

Jardine, M., & Collier Cameron, A. 2019, MNRAS, 482, 2853

Linsky, J. L. 2017, ARA&A, 55, 159

Linsky, J. L., Hunten, D. M., Sowell, R., Glackin, D. L., & Kelch, W. L. 1979, ApJS, 41, 481

Linsky, J. L., Wood, B. E., Youngblood, A., et al. 2020, ApJ, 902, 3

Meunier, N., Lagrange, A. M., Mbemba Kabuiku, L., et al. 2017, A&A, 597, A52

Pasquini, L. 1992, A&A, 266, 347

Sasso, C., Andretta, V., Terranegra, L., & Gomez, M. T. 2017, A&A, 604, A50

Schöfer, P., Jeffers, S. V., Reiners, A., et al. 2019, A&A, 623, A44

Stencel, R. E., Mullan, D. J., Linsky, J. L., Basri, G. S., & Worden, S. P. 1980, ApJS, 44, 383

Thompson, A. P. G., Watson, C. A., de Mooij, E. J. W., & Jess, D. B. 2017, MNRAS, 468, L16

Tilipman, D., Vieytes, M., Linsky, J. L., Buccino, A. P., & France, K. 2021, ApJ, 909, 61

Usoskin, I. G., Solanki, S. K., Krivova, N. A., et al. 2021, A&A, 649, A141

Vernazza, J. E., Avrett, E. H., & Loeser, R. 1981, ApJS, 45, 635

Walkowicz, L. M., & Hawley, S. L. 2009, AJ, 137, 3297

Wise, A. W., Dodson-Robinson, S. E., Bevenour, K., & Provini, A. 2018, AJ, 156, 180

Wood, B. E., Redfield, S., Linsky, J. L., Müller, H.-R., & Zank, G. P. 2005, ApJS, 159, 118

An Introduction to Stellar Magnetic Activity

Gibor Basri

Chapter 4

Transition Regions and Coronae

The transition region lies between the chromosphere and corona and is really just the resulting interface of the very different temperatures and densities that characterize those two regions. This region is both geometrically and optically thin, typically spanning at most a few 100 km. It is misleading to think of it as the thin layer above the chromosphere in a 1-D atmosphere (as depicted in Figure 3.13). More accurately it is the lower skin of coronal loops where the temperature changes rapidly from ten thousand to a million Kelvins. Even the concept of temperature is not fully precise here since particle mean free paths traverse the temperature gradient to various extents and the radiation field is quite complex. Coronal loops are not radial and the length of the loop that experiences transition region temperatures depends on the densities and heating rates in each loop. Finally, the loops are quite dynamic and time-dependent, responding both to changes in their geometry due to footpoint motions in the photosphere and heating changes caused by magnetic reconnection and wave motions.

A conceptual picture of the structures in the solar outer atmosphere is given in Figure 4.1. We will discuss these structures in more detail in Section 4.2 but for now think of the transition region as the bottoms of the various loop structures shown (except for prominences, which are mostly structures at chromospheric temperatures). Observationally the transition region is the region where plasma diagnostics are characteristic of temperatures between chromospheric and coronal. Most of the radiation in transition regions diagnostics arises from the bottoms of the closed coronal loop structures like those in Figure 4.2, although that image is made in the light from a coronal rather than transition region spectral line. Both closed and open loops are visible in the image, as are a couple of cooler denser loops seen in absorption near the bottom of the right-hand side of the main loop structure. This is a side view; looking down from the top (this time in a transition region line) produces images like Figure 4.3. It is good to recall that the chromosphere is also a highly structured region as illustrated by Figure 3.1.

doi:10.1088/2514-3433/ac2956ch4

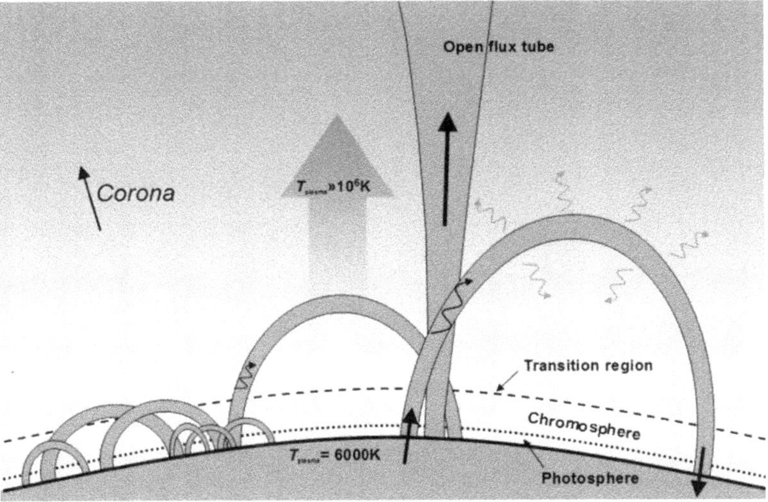

Figure 4.1. A schematic representation of the structures in the solar outer atmosphere. Prominences (low loops on the left) contain cool ($\leqslant 10^4$ K gas) while the closed coronal loops will be more like 2×10^6 K. The open flux tube will be a little cooler ($\sim 10^6$ K) and is the source of the fast solar wind. Sometimes a loop like that on the right can open up at the top producing a helmet streamer and source of the slow solar wind. Credit: This Coronal Loops image has been obtained by the authors from the Wikimedia website, where it is stated to have been released into the public domain. It is included within this article on that basis.

Figure 4.2. An EUV observation of coronal loops with high spatial resolution made by the Transition Region and Coronal Explorer (TRACE) spacecraft. Note that each overall loop is actually composed of many fine threads. The loop bases are brighter, largely because of density. Some loops lie under other loops, but in different directions. The image is taken at 17.1 nm and shows emission from a highly ionized iron spectral line formed at about a million Kelvins. Credit: M. Aschwanden et al. [LMSAL], TRACE, NASA. For a video showing these dynamic loops, see https://svs.gsfc.nasa.gov/cgi-bin/details.cgi?aid=11742. Credit: NASA/SDO, Goddard Space Flight Center.

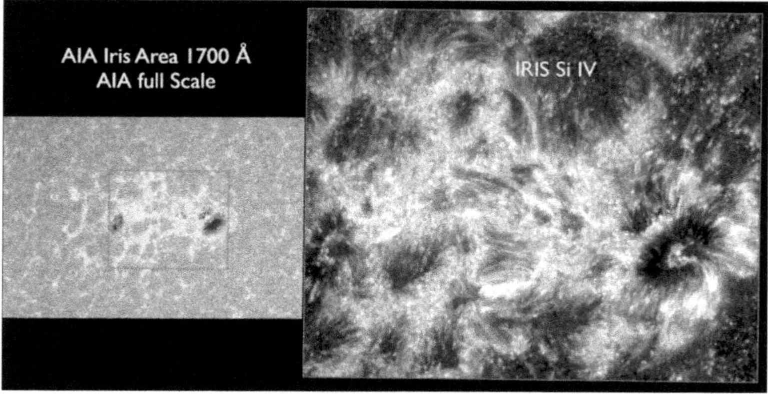

Figure 4.3. UV observations made simultaneously by the IRIS spacecraft of an active region (with different angular scales). The image on the left is taken in light from a silicon bound-free continuum at 170 nm diagnostic of the temperature minimum (4300 K). On the right the image is in light from the Si IV resonance line at 140 nm diagnostic of the transition region (60,000 K). Credit: Reprinted from https://www.universeto-day.com/107136/iris-glimpses-an-elusive-region-of-the-sun/). NASA/IRIS.

The steep temperature gradient arises because of the cooling behavior of solar composition plasmas. Optically thick spectral lines and continua provide radiative cooling to balance non-radiative heating mechanisms in the chromosphere. Radiative cooling is the means by which electron kinetic energy is lost to outgoing radiation through collisional excitation or ionization followed by radiative de-excitation or recombination. The radiative cooling channels operating at the top of the chromosphere are predominantly Lyα and the Lyman continuum of neutral hydrogen (which is ionized at around 10^4 K) and their analogs in helium (which is ionized at around $10^{4.5}$ K). Because both of these species have non-local thermodynamic equilibrium (NLTE) populations, it is a complex calculation to determine exactly how much cooling they supply, but once they are ionized the plasma has no effective way to cool itself until the temperature reaches the regime of 10^6 K. Cooling in the corona is accomplished partly by radiation in free–free continua (primarily bremsstrahlung). The corona also has a more important cooling channel due to conduction down the loops to the denser atmosphere where energy can be radiated away. These thermal balances (which are density and geometry dependent) establish the structure of the transition region. Similar radiative mechanisms cause the interstellar medium (ISM) to have similar characteristic temperature regimes, although with very much lower densities and different heating mechanisms.

4.1 UV, FUV, EUV Spectra

The primary means of observing the transition region is by measuring fluxes (and in a few instances profiles) of spectral lines characteristic of species whose dominant ionization state is in the range of transition region temperatures. Some of the most observed and utilized transition region lines are (listed from lower to higher temperature with the ion, wavelength (nm), and log(T) K): C II (133.5/133.6, 4.3), Si III (120.7, 4.45), Si IV (139.3/140.1, 4.8), C IV (154.8/155.0, 5.0), O VI (140.1, 5.15), and N V (123.9/124.2, 5.2).

Observations of these lines on other stars began in earnest with the advent of the International Ultraviolet Explorer (IUE) satellite in 1979 (Boggess et al. 1978). This 40-cm telescope was in geosynchronous orbit over the Atlantic, with ground stations at Goddard Space Flight Center and Vilspa, Spain. It was one of the few "real-time" space observatories ever flown. Astronomers sat at computer consoles and directed the telescope operator sitting next to them to perform the next observation based on how the last one went, how long a slew was needed to the next target, and what the space weather (radiation fluxes that created noise in the detectors) was like at the time. Even after the pointing gyroscopes began failing, the telescope was utilized in imaginative (if less efficient) two and one gyro modes (presaging the same situation that later occurred with the Kepler mission). Observations of ultraviolet (UV) stellar spectra later continued with the Hubble Space Telescope (HST; on which gyroscopes could be replaced and spectrometers improved), and shorter wavelengths were covered in the 1990 by the Far-Ultraviolet Spectroscopic Explorer (FUSE) and Extreme-Ultraviolet Explorer (EUVE) space telescopes described below. An historical review of UV astronomy was written by Linsky (2018).

A set of exemplary spectra at different wavelengths and resolutions are presented in Figures 4.4, 4.5, and 4.6. The spectra include some chromospheric lines and a few coronal lines along with those from the transition region. Figure 4.4 shows a calibrated low resolution UV spectrum of the active Sun so the shape of the underlying continuum is also apparent. Segments of those wavelengths and a continuation up to the Mg II lines at high resolution are shown in Figure 4.5. Such spectra are sufficiently resolved to reveal Doppler motions in the lines. The transition region is revealed to be non-thermally broadened (with sometimes highly

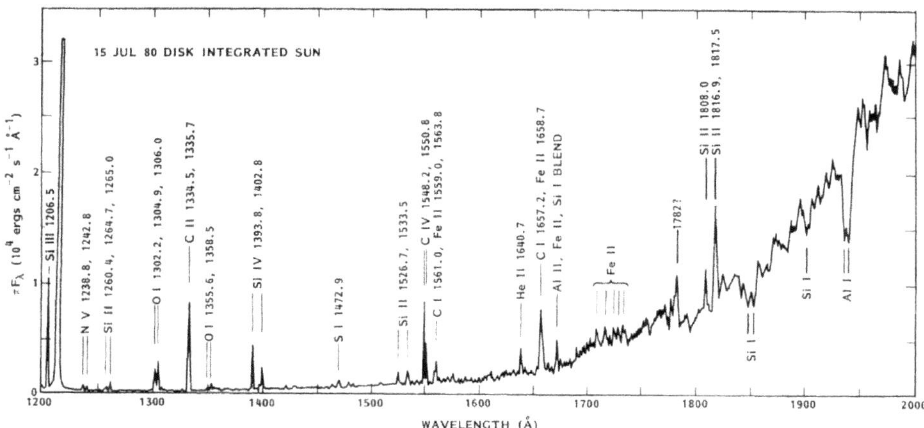

Figure 4.4. A calibrated lower resolution UV spectrum of the active Sun, taken by a rocket. The very bright line on the left is Lyα. Credit: Mount & Rottman (1981), John Wiley & Sons. Copyright © 1981 by the American Geophysical Union.

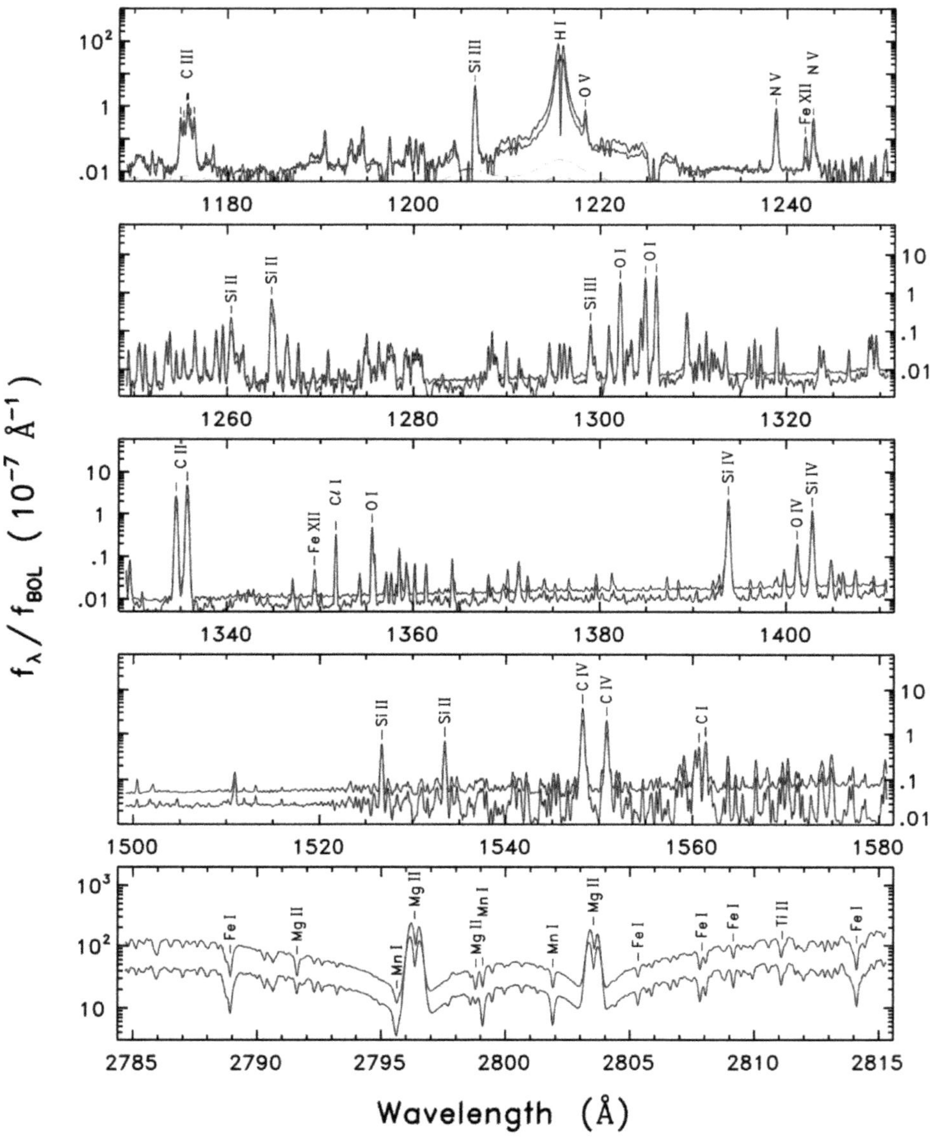

Figure 4.5. High resolution UV spectra taken by HST with the Space Telescope Imaging Spectrograph (STIS) spectrometer of the nearest solar-type star α Cen A (5800 K, 1.1 $M_\odot$) in blue, and its companion α Cen B (5230 K, 0.94 $M_\odot$) in red. Many lines from the transition region are seen in emission (along with the chromospheric Mg II and H I resonance lines). Credit: Reproduced from Ayres (2020).

supersonic turbulence) and sometimes shows a blueshift. The instrumental challenge (in both optics and detectors) grows more difficult when trying to get to shorter far ultraviolet (FUV) wavelengths. The primary stellar observatory at these wavelengths so far has been the FUSE spacecraft (Moos et al. 2000). Figure 4.6 shows the

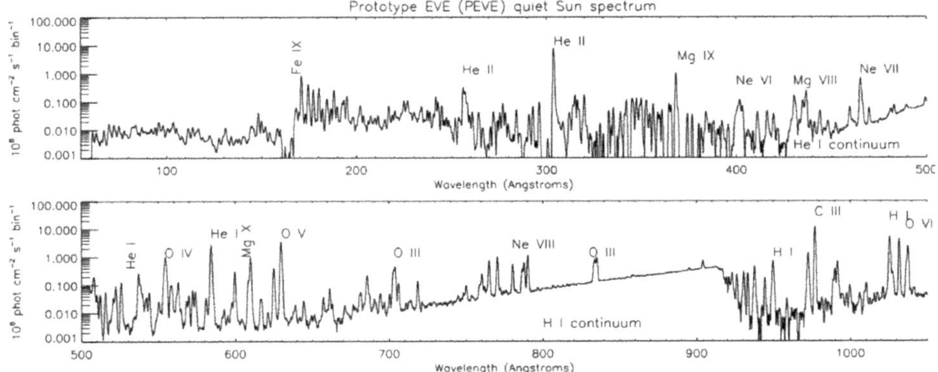

Figure 4.6. The FUV/EUV spectrum of the Sun, taken by the EVE (Extreme Ultraviolet Variability Experiment) spacecraft. The Lyman continuum is the very prominent sloped section shortward of 912 Å but this region is generally wiped out for other stars due to the ISM. Credit: Reprinted by permission from Springer Nature: Del Zanna & Mason (2018), © 2018, The Authors.

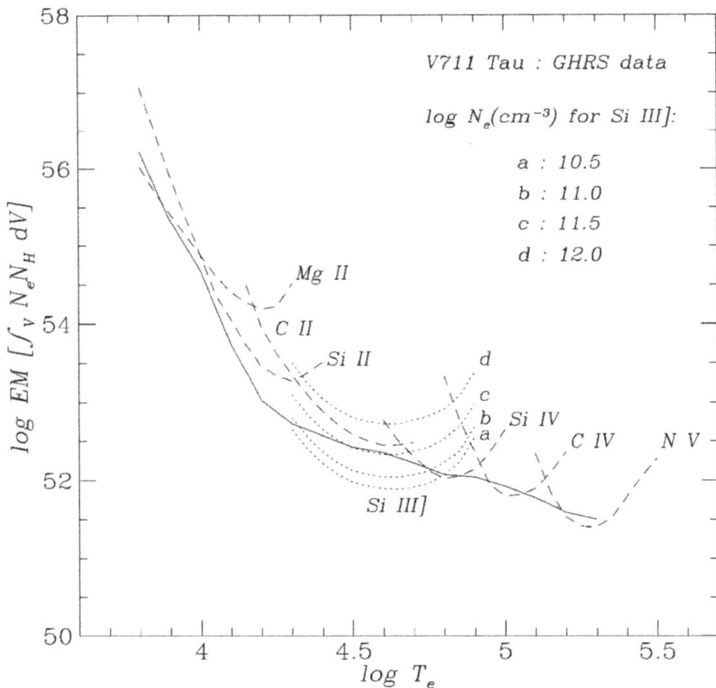

Figure 4.7. An example of an emission measure analysis. The target is an RS CVn system (very active close binary system) and observations were made with the Goddard High Resolution Spectrograph (GHRS). The atmosphere is taken to populate a locus (solid line) set along the minima of the curves or by using a more sophisticated combination technique. The dotted lines show how the assumed electron density can change the emission measure for a particular line; inversion of this yields a preferred electron density. Credit: Reproduced from Griffiths & Jordan (1998). © 1998. The American Astronomical Society. All rights reserved

solar spectrum all the way down to near X-ray wavelengths. The region of the Lyman continuum (90 nm) and below is called the extreme ultraviolet (EUV); observing it presents a major problem for all stars but the Sun because the interstellar medium (ISM) is generally quite opaque at those wavelengths. Luckily the Sun is in a local (few parsec) ionized bubble, and the neutral ISM beyond is patchy enough so that some stellar observations (Craig et al. 1997) have been obtained shortward of 50 nm by the EUVE spacecraft (Bowyer et al. 1994); it even obtained a few extragalactic observations. The grazing incidence optics needed for this wavelength region are like those required for X-rays.

The temperature and density of formation of these species is the subject of much work in laboratory experiments and plasma codes. Those results have been summarized and made usable in codes and databases such as CHIANTI (http://www.chiantidatabase.org) that are extensively employed in the analysis of optically thin lines at short wavelengths from hot plasmas. Because the lines are optically thin (so we can ignore self-absorption and opacity) and formed against essentially zero background (either empty space or a much cooler photosphere) it is straightforward to write down the observed intensity assuming that the source function consists solely of spontaneous emission (see Equation (A.4)):

$$F_{\nu}(0) = \int_0^{\infty} S_{\nu}(s)ds \quad \text{where} \quad S_{\nu}(s) = (h\nu/4\pi)n_u(s)A_{ul}. \tag{4.1}$$

One often then makes the equilibrium "coronal approximation," namely that $n_u(s)A_{ul} = n_l(s)n_eC_{lu}(T_e)$ and that the electron density and temperature can be treated as constant at appropriate values. This assumes that the upper state (which is the source of emitted photons) is only populated by collisional excitation of the lower state by electrons. Of course, only electrons with energies above the minimum requirement can do that (and that population depends exponentially on temperature). There is also an implicit assumption that the population of ions producing the transition being considered is set by ionization equilibrium processes that are also driven by electron collisions (ignoring photoionization because the species in question require high-energy radiation to ionize them). There will therefore be a preferred set of temperatures at which the chosen ion is dominant in the population of the element in question, and the chosen line is thus excited. The path integral can therefore be viewed as an integral in temperature and density, and the emergent intensity re-written as:

$$I_{\nu}^0 \propto \left[\int_{T_1}^{T_2} G_{lu}^1 \xi(T_e)dT_e \right] \propto \left[G_{lu}^2 n_e n_H \xi(T_e)dV \right]. \tag{4.2}$$

The quantity $\xi(T_e)$ is called the "emission measure," and it is a function of n_e^2 or equivalently $n_e n_H$ (assuming that hydrogen is fully ionized) and T_e. There are a number of other implicit assumptions contained in G_{lu} (which takes different forms in different methods; the computation of emission measures is something of an art). More detailed and extensive discussions of emission measures can be found (for example) in chapter 5 by Judge in Engvold et al. (2019) and Griffiths & Jordan (1998). One question is what the interval between T_1 and T_2 (where the ion in question is dominant) should be. A choice that has been used is 0.3 in $\log(T)$ when the electron density, temperature gradient, or pressure are taken as constant over the region of line formation. More detailed calculations can take account of the full contribution function and gradients in these variables. The integral really stands in for a volume integral over the part of the atmosphere where the spectral line is formed, which can be done in plane-parallel geometry, spherical geometry, loop geometry or something else. The truth is that we don't know what the configuration really is on other stars.

The coronal approximation itself can be replaced by something more realistic by including a number of additional physical processes including NLTE effects, treatment of more detailed atomic physics like dielectronic recombination, potential optical depth effects, and non-Maxwellian electron distributions. The use of emission measures is thus best done as a comparative exercise of different stars with similar assumptions. An example of a paper studying stellar activity using the full set of spectral lines from X-rays to the UV is Ness & Jordan (2008). The primary purpose of such exercises is to assess the amount and density of plasma at different temperatures in coronal loops. The emission measure grows with both the volume of plasma at a given temperature and also its luminosity per unit volume, and the temperature and density distribution can vary along a loop (or in other geometries).

One very convenient observational result is that the fluxes in the various transition region UV lines are quite correlated with each other. The slopes of these correlations vary somewhat, tending to increase with the temperature of formation compared against the same cooler diagnostic and steepest for X-rays. It is possible to predict with an accuracy that is consistent with the intrinsic scatter of the fluxes at a given level of activity what the fluxes of all the lines will be if one has measured one of them well. An illustrative summary of these observations is given by Oranje (1986) and shown in Figure 4.8. A more recent paper (France et al. 2016) shows that one can also predict the general UV and FUV fluxes from a star if one has measured, for example, C IV (or Mg II). The ability to use more easily observed diagnostics is important if one wants to predict unobserved FUV/EUV fluxes from stars onto exoplanets for use in evaluating their effect on planetary atmospheres.

There has not yet been a simple explanation of why the correlations are so good, although given the constricted nature of the transition region and the averaging over the whole star it perhaps is not too surprising. One of the reasons Figure 4.8 looks so good is its logarithmic scale. A few studies have found possible (but not large) slope changes between various pairs of lines or various types of stars. It is remarkable nonetheless that these relations hold over stars with very different effective temperatures, surface gravities, rotation periods, and ages. They speak to a commonality in how magnetic heating affects a stellar atmosphere overall despite the great complexity of structures that are generated. Of course the line fluxes from a given star are variable over time and change even more substantially during flares.

4.2 Coronae

The nature of the solar corona was not understood until nearly halfway through the 20th century. As mentioned in the Introduction the visible light seen in eclipses is due

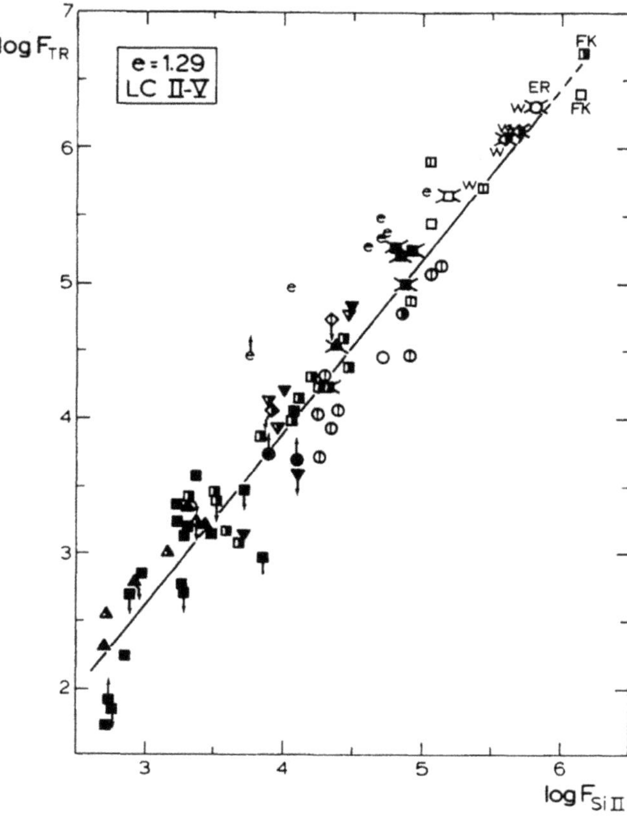

Figure 4.8. The relation between the logarithmic fluxes in the chromospheric Si II (126.5 nm) lines compared with the sum of the transition region lines Si IV, C IV, and NV, as observed by IUE. The circles are main sequence stars and downward triangles are supergiants. The emptier the symbol, the hotter the star. The "e" symbols denote dMe stars and the labeled stars at the upper right are RS CVn binaries. They all lie on essentially the same power law relation. Credit: Oranje (1986), reproduced with permission © ESO.

to Thompson scattering of photospheric light; free electrons are abundant in the highly ionized corona. Interestingly, a visible spectrum of the corona also includes a green emission line at 530.3 nm, whose identification with Fe XIV in the 1930 s was one of the early clues that the corona is very hot. Spectroscopic data on highly ionized species at millions of degrees had just become available. The idea that the solar outer atmosphere could be so hot was far from obvious, even far-fetched. The primary X-ray emission from coronal plasma could not be seen until rockets got above our atmosphere. It is due to bremsstrahlung and other processes that finally provide enough radiative cooling to balance the heating and stabilize the temperature at millions of degrees.

Plasma in the loops is being heated by mechanisms that are not fully understood and are being investigated more deeply by several solar space missions at the time of writing. It is thought that they must involve some combination of MHD wave

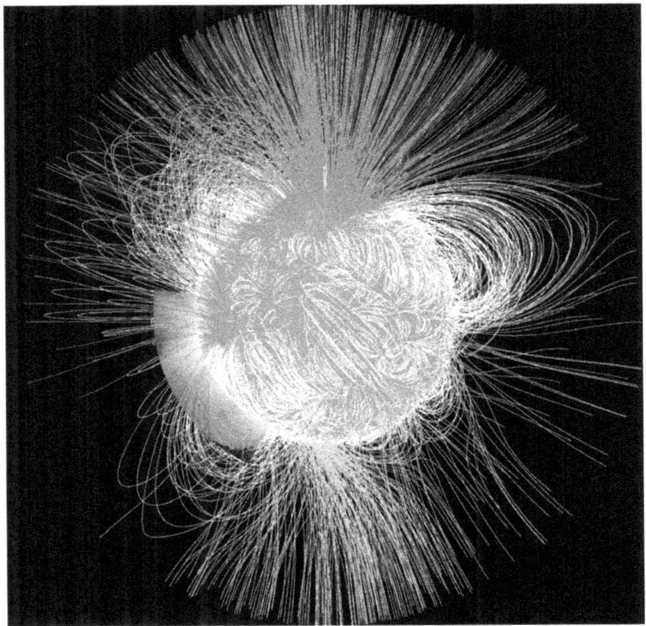

Figure 4.9. The structure of the solar coronal magnetic field inferred from photospheric magnetograms and EUV images of coronal emission. The field is assumed to have taken a potential (stress-free) configuration, and open fields (both polarities) are colored. Credit: © UCAR.

dissipation, resistive currents, and magnetic field reconnection mechanisms (like micro-flaring). Recent simulations model the field from inside the photosphere at one footpoint of magnetic concentration through the loop to the other footpoint and include photospheric convective motions (Janvier et al. 2015). The model loop is composed of thin (few hundred km) strands of similar field lines (cf Figure 4.2). The strands move into contact with each other and sometimes twist around each other. This generates a lot of small-scale reconnection currents and energy travels up the strands and heats the loop. Wave motions can also be excited and play a role. The heating is dynamic on the scale of a few minutes or less. This literature is likely to become rapidly more extensive as the current solar missions collect new data.

The emission is brightest from closed magnetic field loops. The densities in these loops are highest for those rooted in active regions and their emissivity increases like the density squared, so the more field in the active region the brighter the coronal loops are. The loops are dynamic in both time and space; turbulent motions in the photosphere and re-arrangements of magnetic field cause ever-changing configurations. Sometimes this leads to massive explosions due to sudden dissipation of magnetic fields, called "flares" (Section 4.3). The primary outputs from these are extremely hot and high-energy radiation (X-rays and γ-rays). They also accelerate particles that stream down and heat the chromosphere and photo-sphere (creating visible and UV radiation) and/or blast off the star carrying a lot of mass and energy.

The overall magnetic structure of the corona is often modeled as a set of potential field configurations rooted in bipolar regions on the solar surface. For the Sun, the field on the surface can be known in great detail using magnetograms. These give the field strength and polarity with good spatial resolution, and theory can then be used to infer the vertical structures above the photosphere by assuming they are in their most relaxed (potential) configuration. These calculations can be augmented by observing the actual coronal loops in high-energy radiation. For other stars, Zeeman Doppler Imaging provides a slightly spatially-resolved version of the stellar field (or at least the field left over after sub-resolution canceling of opposite polarities). The same potential field theory can be applied to give an idea of at least what the field not too close to the stellar surface is, since smaller closed loops will not extend too far above the star. We return to this subject in Section 6.2.2.

A variety of coronal structures are observed on the Sun (Figure 4.1). There is a general dipole component to the solar field that manifests as a set of open field lines near the poles. The field over active regions tends to be closed, although the loop arcade can be open near the top, creating a helmet streamer. The white-light corona tends to be more symmetric during solar minimum and have equatorial extensions during solar maximum (which extend many solar radii in extremely high contrast images). Of course, the photosphere is dynamic and the loop footpoints move around, forcing stresses on the actual field. These stresses get relieved by a number of dynamic mechanisms, the most dramatic of which are flares and coronal mass ejections.

The average loop density structures are not in hydrostatic equilibrium (there is more density at the tops of the loops than would be expected). Simple loop models have been developed to help diagnose basic physical conditions from observations. A well-utilized example of these is from Rosner et al. (1978). The loop equilibrium is characterized by scaling laws that connect the energy flux and loop length to the resulting temperature and density structure. Temperature and density are assumed to be co-dependent, and the main mechanism of heat redistribution is conduction in the loop which is very efficient in carrying energy back out of the loop. Making a number of simplifying assumptions (uniform heating, constant loop cross-section and others) they derive a very simple expression relating the loop maximum temperature to its length and density: $T_m = 1400(L\rho)^{1/3}$ in cgs units. Variants of this approach dropping different assumptions have been calculated by several other authors.

These loop models can and have been applied to X-ray observations of other stars; they are most useful in a differential comparison between stars. A variety of observations of stellar coronae began with the High Energy Astrophysics Observatory (HEAO)-1 and much more influential Einstein (HEAO-2) spacecraft (1977) followed by the European X-ray Observatory Satellite (EXOSAT; 1983). These missions provided X-ray luminosities and crude X-ray spectra of hundreds of nearby active stars. A foundational review can be found in Rosner et al. 1985; an updated review was done by Güdel & Nazé (2009). The early missions were followed by the Roentgen Satellite (ROSAT; 1990) which performed an all-sky survey, the Advanced Satellite for Cosmology and Astrophysics (ASCA; 1993) which was the first to use CCDs for X-rays, then the major X-ray observatories CHANDRA and Newton-XMM

(X-ray Multi-mirror Mission; both in 1999) that have conducted thousands of deep pointed observations and provided resolved X-ray spectra well into the 21st century. At the time of writing, the extended Roentgen Survey with an Imaging Telescope Array (eROSITA; 2019) mission is conducting a new all-sky X-ray survey that is 30 times deeper than ROSAT. With such data one can compare the total X-ray luminosities of stars as a function of their mass, age, radius, or rotation period to see how stellar coronae are related to these parameters. eROSITA also has spectroscopic capabilities similar to XMM-Newton. Measurements of coronal temperatures will allow conversion of luminosities to emission measures or integrated loop properties.

Once even very low resolution X-ray spectra were possible, it became clear that more active stars appear to have an extended range of coronal temperatures. This is sometimes interpreted as the presence of a roughly solar-like component of a few million degrees and a hotter component of 10–25 million degrees. The more active the star, the more of the hotter component is seen. An example of a two-component loop analysis can be found in Schrijver et al. (1989). As better data was obtained it often looks like there is a range of temperatures present (Figure 4.10). Because there is no spatial resolution, this can be interpreted as a range of loops with different densities and temperatures. A detailed analysis of EUV and X-ray spectra for a very bright source (Capella) can be found in Brickhouse et al. (2000). Because the overall corona is relatively space-filling and the cooling rates don't vary strongly with temperature, the variation in X-ray luminosities can be interpreted as primarily due to a variation in densities within loops of different temperatures. The solar loops are certainly brighter where their densities are higher over active regions. The full range of stellar coronal properties cannot be explained by changing the filling factor of

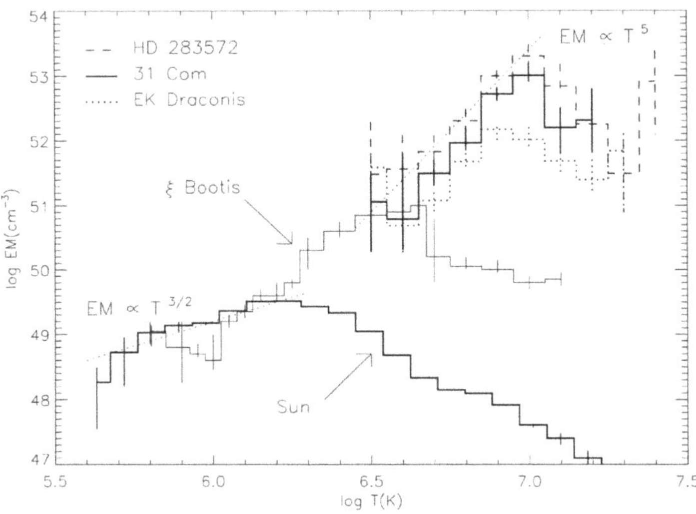

Figure 4.10. X-ray volume emission measures for a set of stars increasing in activity from the Sun. These were taken with various spacecraft and also represent a variety of stellar types from very young to evolved off the main sequence. Only the constrained portions of the emission measures are shown. Credit: Scelsi et al. (2005), reproduced with permission © ESO.

solar active regions, however. It must be the case that some stars produce loops that are brighter (larger, denser) than any solar loops. There is evidence on red dwarfs that some loops can span much of the stellar radius (unlike the Sun), and RS CVn stars show evidence of loops that extend between the binary components (e.g., Walter et al. 1983; cf Section 7.3).

Stellar X-ray luminosities are significantly more variable than optical luminosities. This is not surprising given the steep power law relations between low and high temperature diagnostics, and the fact that the Sun exhibits such behavior. A survey of the Hyades cluster by ROSAT (Stern et al. 1995) showed this variability, although the most active stars are less variable. This is presumably because they are always very active; the activity may be "saturated." A deep CHANDRA pointing on the Orion star-forming region also shows a lot of variability (Stassun 2007) and the variability appears to arise from magnetic activity rather than accretion on those very young stars. One has to be careful when looking at star-forming regions because accretion can produce luminosity changes through physically different processes. Of course, flares also cause short-term variability that can have large amplitudes across the entire electromagnetic spectrum. Optical flares are generally also seen in X-rays but X-ray flares do not always produce optical flares. Pye et al. (2015) provide one survey of stellar X-ray flaring activity for an unbiased sample of hundreds of stars (although flares are detected on only about a tenth of them). They don't find a large difference between the mean X-ray luminosities of stars that flared and those that didn't, but caution that the statistics in their survey are not definitive. What we know for sure is that the red dwarfs known as "flare stars" have high flare rates compared with other red dwarfs (Hawley et al. 2014) and are young and rapidly rotating.

In the early days of stellar X-ray observations it appeared that there could be a dividing line in the HR diagram located in the Hertzsprung Gap (between the main sequence and the giant branch) on the cooler side of which coronal X-ray emission was not seen. As observations continued and gained sensitivity, "hybrid" stars were found which show evidence of transition region (and sometimes coronal) plasma although at lower levels compared with their chromospheric activity. In a later and detailed study of this issue Ayres et al. (1998) conclude that transition region and coronal plasma can be present even on the giant branch but that X-ray luminosities are generally much lower relative to lower energy diagnostics. It appears that some hot gas is present but the bulk of the gas is increasingly cooler. We return to the possible explanation of this in Section 4.4.

Another part of the main sequence where X-ray emission greatly weakens is the A stars; they lie between stars with outer convective envelopes like the Sun and hot stars with strong radiatively-driven winds. Although some A stars are observed to have X-rays it is quite possible they are emitted from unseen low-mass companions. O and B stars exhibit considerable X-ray emission; it is not due to coronae but rather to shocks being driven by clumps within their radiatively-driven winds that are more rapidly accelerated (Section 7.4). The other part of the HR diagram where coronal emission really disappears is below the bottom of the main sequence (Section 7.1).

Stellar coronae are also observed at radio wavelengths. Incoherent (broadband) radio emission is caused primarily by electrons, either accelerated by encounters with ions (bremsstrahlung) or when spiraling through magnetic fields (synchrotron). In addition, the electrons can interact with resonant plasma waves and produce coherent, polarized emission of very high brightness temperature. A foundational review of this topic was given by Dulk (1985) and a more recent one by Güdel (2002). In addition to continuum radiation from the general magnetic atmosphere, there are various types of radio bursts caused by flares (Section 4.3) and coronal mass ejections (CMEs; Section 4.4). Some are caused by blasts of electrons creating shock waves in the corona, others are due to plasma waves associated with magnetic reconnection and other causes. These are easily observed on the Sun, but it requires intrinsically high luminosities to see them on other stars. As radio observations have gotten more sensitive, an increasing number of stars have been detected in quiet coronal radio emission. Atacama Large Millimeter Array (ALMA) can see them at sub-millimeter wavelengths and the extended Very Large Array (eVLA) and other radio observatories at mm and cm wavelengths. More active stars have brighter continua and produce larger and more frequent flares. Polarized emission from flares can be seen on all the active types of stars. Even brown dwarfs sometimes produce radio emission; in that case the radio emission is very strong relative to the X-ray emission (the latter is mostly missing). We return to this topic in Section 7.1.

4.3 Flares

Wherever there are strong magnetic fields accompanied by plasma flows the field can find itself moved into stressed configurations that can be reduced to a lower energy state by reconnecting field lines into a simpler form. This magnetic reconnection releases the excess energy through current sheets that can also accelerate charged particles to very high energies. The energy release occurs on timescales ranging from milliseconds to hours depending on how much reconnection is taking place over what volume and by what mechanism. Some of the energy goes into high-energy photons or relativistic sprays of particles, some of it goes into heating of the local plasma by these effects, and some can go into kinetic motion of large volumes of plasma directed by the reconfiguring magnetic fields. This general type of energy release is called a "flare."

Flare energies on stars can range from hard-to-measure "nanoflaring" to solar flares that lie on a power law (with index about -2) up to 10^{34} ergs in electromagnetic radiation for the largest (they can also release a significant amount of kinetic energy). A typical large flare on the Sun might emit 10^{33} ergs. "Superflares" have been observed on solar-type stars with energies up to about 10^{36} ergs, and enormous flares on T Tauri stars or low-mass stars that can total up to 10^{38} ergs have been observed. These monster releases can cause the total optical luminosity of a red dwarf to temporarily increase by more than an order of magnitude in extreme cases. Good general reviews of flares have been written by Hudson & Ryan (1995) & Benz (2010) among others.

The main question about the flare mechanism is how the magnetic field can dissipate energy as quickly as it does while producing very efficient high-energy acceleration of electrons and protons. Reconnection can occur if magnetic field lines with opposite polarity come into contact with each other. This requires that plasma motions introduce sufficient stress into the field configuration since otherwise that would not happen in relaxed or potential field configurations. The field then relaxes to a simpler configuration in which the polarities are again separated; this requires the destruction of some field lines by releasing their magnetic energy. The issue then comes down to what sort of mechanism on micro-scales can convert magnetic field lines to energy very efficiently and what configuration of flows can move the energized material away fast enough and keep bringing in new field at a sustained rate.

The conceptual model of reconnection that has been dominant since it was proposed is the Petschek mechanism (Petschek 1964) illustrated in Figure 4.11 from Lee & Lee (2020). The original reference is only a conference report that also contains a short interesting discussion afterwards with Drs Parker and Sweet, who proposed the original flare ideas. What exactly physically happens in the reconnection region is still not fully understood. Ohmic diffusion is vastly too slow and various proposed collisionally-based anomalous diffusion mechanisms are much faster but still generally too slow to explain what is observed. When the diffusion is collisionless (meaning that the field is converted on scales smaller than the mean free paths of the particles) the electron diffusion region becomes smaller than the ion diffusion region, and the reconnection rate speeds up by orders of magnitude. One recent discussion of the state of affairs in theory is provided by Liu et al. (2017), who provide conceptual hope of a solution but not a final detailed model. Modern theory and simulations of the actual reconnection processes, especially those that extend to three dimensions (which introduces a richer set of scenarios) seem able to operate at the needed rate but a full model is not yet accepted. Magnetic reconnection is actually responsible for large releases of energy and high-energy particles in a variety of astrophysical contexts so this topic receives a lot of observational and theoretical attention, but the details are obviously tricky.

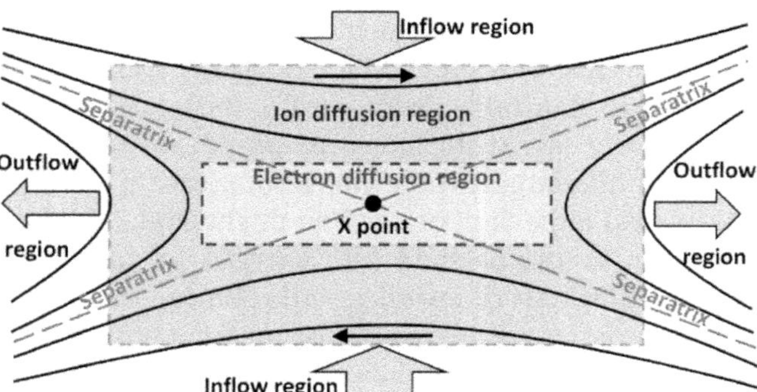

Figure 4.11. The proposed basic configuration (Petschek mechanism) of magnetic reconnection in flare situations. Field lines pointing to the right are brought in from above and contact field lines pointing to the left coming in from below. Collisionless diffusion occurs in the ion and electron diffusion regions, converting the opposite field lines into energy that causes outflows (slow mode shocks) to the right and left, creating space for the next field lines to move in and reconnect. Credit: Reprinted by permission from Springer Nature: Lee & Lee (2020), © 2020

There can be a variety of magnetic configurations that lead to the reconnection that powers flares on different scales. On the smallest scales, plasma turbulence can cause very local twists of the field so that opposite polarities come into contact. Emerging flux loops can run into regions of opposite polarity above some parts of them. Loops can be carried into conflicting contact with other loops by convective motions in the photosphere through twists or shear. Parts of a loop can come into conflict with each other because the loop is being distorted by footpoint motions. Flares are often found to be initiated near the tops of loops as illustrated in Figure 4.12, which shows the canonical conception of how flares occur. In that picture a plasmoid can be ejected upward (something like a CME) while much of the outflow from the reconnection region moves down the newly simplified lower loop and heats the chromosphere (and

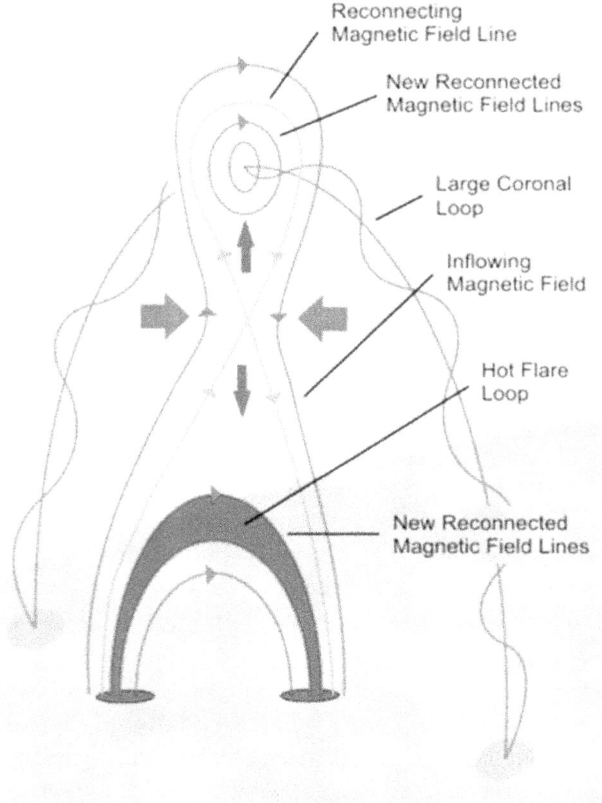

Figure 4.12. The typical conception of a solar/stellar flare. Reconnection itself occurs at the "X-point" shown in the middle where opposite field polarities come into contact. New field is brought in as indicated by the thick blue arrows to replace the plasma that leaves along the thinner red arrows. A closed field plasmoid can be ejected above, while the simplified loop below directs energized plasma down to its footpoints. The situation may have been initiated by a rising loop coming into contact with the large coronal loop also shown, which may also have a helical structure due to shears at its footpoints. Credit: Reprinted with permission from NASA.

photosphere if strong enough). High-energy particles are launched in both directions. The reconnection region itself is very energetic and radiates X-rays and γ-rays.

Our observational understanding of high-energy processes in solar flares has been greatly improved by a fleet of spacecraft over the last 30 years. These include a series of Geostationary Operational Environmental Satellite (GOES) satellites that had X-ray detectors. Solar flares are given a letter classification based on the X-ray energy received by GOES (or its equivalent): A, B, C, M, and X in order of logarithmic energies (W m^{-2}) from less than -7 (A) to greater than -4 (X, which can have numbers appended to indicate how energetic). A C class flare is apparent in Figure 3.1. The Compton and Reuven Ramaty High Energy Solar Spectroscopic Imager (RHESSI) spacecraft measured γ-ray fluxes. Imaging in the EUV and X-rays has been conducted by later versions of GOES, Yohkoh, Hinode, Solar Maximum Mission (SMM), TRACE, SOHO, Solar Dynamics Observatory (SDO), IRIS, and in γ-ray by RHESSI. Further progress is expected from missions that also approach closer to and/or over the Sun including Solar Orbiter (out of the ecliptic) and the Parker Solar Probe (closer); Japan is also planning an EUV mission called Solar-C for now.

There is an extensive literature of phenomenology of flares observed in all the relevant diagnostics with increasing spatial resolution. One example of a flare observed in a number of diagnostics simultaneously is provided by Lee et al. (2017). These all contribute to the general picture shown in Figure 4.12 but they also show that there is a wide variety of flare types and evolution and phenomenologies. It is less clear now that large flares are mostly confined to the tops of loops as there is sometimes simultaneous emission from lower down into the transition region. The Solar Orbiter has recently seen direct evidence of micro- and nano-flares (these refer to fractions of the canonical 10^{33} ergs) in the chromosphere and corona. The small low-lying ones were dubbed "campfires;" they last 10–200 s and are about the size of granules and can range down to EUV energies of only 10^{23} ergs. There is a growing suspicion that reconnection on this scale may provide the dominant mode of coronal heating, both directly and also indirectly through waves it excites. One line of evidence that has been around for some time is that coronal X-ray fluxes show a constant small jitter. Not much is currently known about the mechanisms and modes of these small flares. In general flare energies for a particular star seem to follow a power law: $\frac{df(E)}{dE} \sim f_0 E^{-\alpha}$ where $\alpha \sim 1.5$–2.5. It is harder to assess the small energy tail of flares on stars because of sensitivity problems, but there has to be a low energy cutoff to keep the integral finite. There seems to be rather little evolution of the power law spectrum with age for a given type of main sequence star.

Larger flares unfold on various timescales, generally the higher the energies involved within a flare the shorter the timescales, but the larger in volume the flare the longer the timescales. The latter point is because an increasing number of loop threads or adjacent loops may get involved. As the initial high-energy particle accelerations take place, radiation is produced at similarly high energies by bremsstrahlung and inverse Compton scattering. A variety of particles can also be produced that decay (like pions) and various nuclear transitions can also be excited. All these processes produce hard X-rays and γ-rays. The composite radiation ranges from 10 keV to many MeV as a power law with steep indices (3–4) that become shallower over time. There is accompanying coherent and incoherent radio emission as particles move within the fields in bulk and singly, and sometimes sprays of Hα emitting plasma that may be more directly associated with the reconnection jets. There is a lot of non-thermal (accelerated particle) plasma, but also a thermal component that can have temperatures ranging from 10–100 or more millions of Kelvin. Solar plasma has been observed explosively expanding at several hundred km/s in response to the sudden deposition of energy. This initial "impulsive phase" usually lasts for timescales of seconds to minutes, with much faster variability (as short as tens of milliseconds) seen within it. Consult Benz (2010) and Janvier et al. (2015) for more detailed discussions of the many possible acceleration and excitation processes under study. Most of them involve some mix of direct currents, shocks, turbulence, or interactions with the magnetic field like Fermi acceleration.

During the later part of the impulsive phase the following gradual phase begins but then continues for far longer (many minutes to hours). In this phase the some of the initial energy and particle flows released travel down toward denser plasma and begin heating it. Direct particle beams (both electrons and protons) can travel down the magnetic loop and evaporate plasma at the footpoints through the transition region into the chromosphere and sometimes into the photosphere. Radiation can also spread more broadly toward the surface of the star. More non-thermal energy is released by flares than thermal energy (by a factor of a few). Once struck, the chromosphere begins to heat up and re-radiate the radiation as soft X-rays, UV and optical (and IR) light and plasma can also begin to evaporate and flow upward. The radiation can be optically thick and produce a blue/UV continuum as well as greatly exciting and broadening chromospheric spectral lines. There is often a pre-heating phase of a few minutes before evaporation starts in earnest. The evaporation can occur either "gently" or explosively depending on how quickly energy is deposited and whether the overpressure can be accommodated by thermal expansion or not. In the case of gentle evaporation, plasma is ejected upward at a few tens of km/s and thermal/turbulent line broadening is apparent. The densities and emission measures in the loop are strongly increased by the new plasma. These effects can be observed in stellar flares as well.

The length of the gradual phase depends in part on how large the flare is; flares need not be confined to one or a few loops but can sometimes trigger a whole loop arcade to light up sequentially (Figure 4.13). On red dwarf stars the flaring loops can become nearly as large as the star itself. In these cases the gradual phase can go on for many hours. There is a tight power law correlation between the X-ray and radio luminosities of flares sometimes known as the Güdel–Benz relation (cf Benz 2010) that covers more than nine orders of magnitude for flares of a large range in strengths on the Sun and other stars (except the coolest stars). This is because the radio emission is caused by the same non-thermal electrons that produce the bulk of the X-ray emission and the two forms of energy together seem to be a signpost of the typical flare mechanism. There is also a good correlation in total energy between the early hard X-rays and later soft X-rays because they both reflect the amount of energy released by reconnection (although in different ways at different times). The

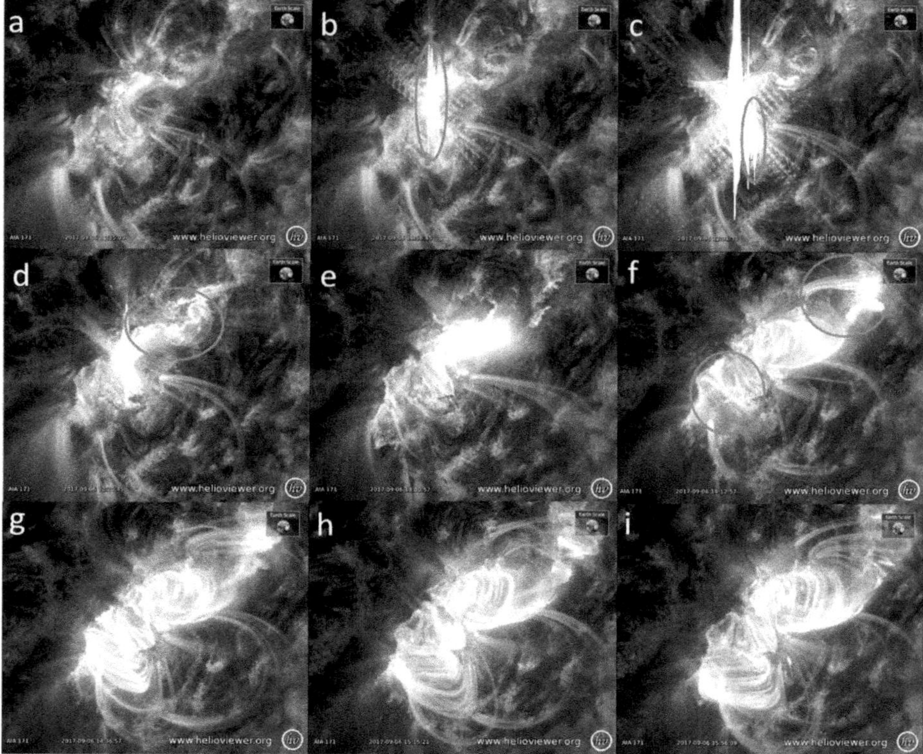

Figure 4.13. An SDO AIA 17.1 nm image sequence during the X9.3 flare on 2017 September 6. Red circles show the location of new postflare loops and are associated with new peaks in the SDO EUV time series. They indicate new regions of reconnection or temporally disconnected regions along the flare arcade that are delayed from the initial, centrally located eruption. Credit: Chamberlin et al. (2018). John Wiley & Sons. ©2018. American Geophysical Union. All Rights Reserved. For a video including flares, see https://svs.gsfc.nasa.gov/cgi-bin/details.cgi?aid=11742. Credit: NASA/SDO, Goddard Space Flight Center.

length of the gradual phase depends on the cooling times in the plasma that gets heated, which are density dependent. Electron densities often seem to lie in the 10^{10}–10^{12}cm^{-3} range. The Güdel–Benz relation is violated for very low-mass stars and brown dwarfs, which have a lot more power directed to coherent radio emission and much weaker X-ray coronae in general (Section 7.1).

Much of the early information on stellar flares came from observations of the Balmer lines and optical photometry (especially in the U & B bands). All the chromospheric lines brighten during flares, and both the Paschen and Balmer continua can also brighten. The ratios of brightening in various diagnostics often resemble solar ratios. Once stellar X-ray observations became possible these correlations extended to high energies. One difference in flares in more active stars is that the flare loops seem to be larger and the flares can occur at greater altitudes (in stellar radii). They also tend to have larger implied loop densities. A surprising discovery during the Kepler mission was that solar-type (5100–6000 K) main sequence stars can exhibit easily-seen

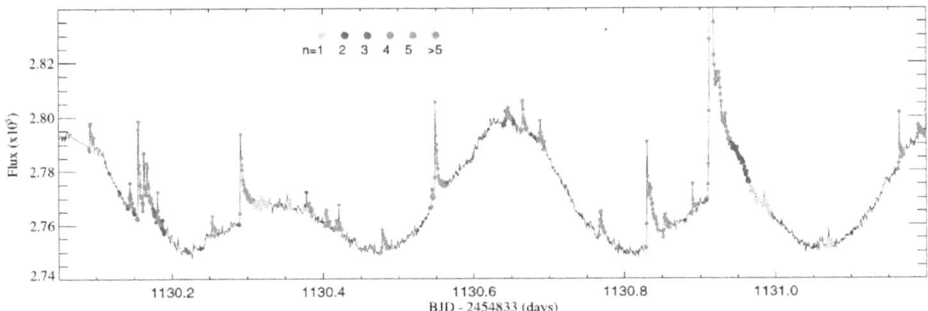

Figure 4.14. The Kepler light curve of the very active flare star GJ 1243 (M4). This is a short cadence observation with a time resolution of about 1 min. The colors indicate a confidence level in the reality of a flare (red is best). In addition to large multiple flares and many smaller ones, the overall slow change in the amplitude of the light curve results from starspots rotating in and out of view. Credit: Reproduced from Davenport et al. (2014). © 2014. The American Astronomical Society. All rights reserved.

white-light flares against the whole star. These stellar "superflares" have energies above 10^{33} ergs. The superflare rate of occurrence increases as the stellar rotation period decreases; younger stars have superflares more often. An estimate from superflare statistical occurrence rates is that the Sun might produce a flare with energy of 5×10^{33} ergs less than once per thousand years (Okamoto et al. 2021). Normal solar flares would only be detectable by Kepler once a century or so because their white-light energies are so much less than the solar luminosity.

There are also sometimes large flares between active close binary stars (particularly RS CVn or Algol systems); these have been spatially resolved by radio interferometry in a few cases. As age is decreased to the pre-main sequence the magnetic flux on the stellar surface can be up to a thousand times greater than that of the current Sun, and magnetic interactions between the star and its disk are also possible in the T Tauri phase. The largest flares seen have been in such systems. Large flares can be very frequent on "flare stars" (usually young M dwarfs) that also have large filling factors of strong magnetic fields (Section 6.2). An example of white-light flares that come every few hours is shown on one of these stars in Figure 4.14. Flare loops tend to cover a larger fraction of the (smaller) stellar surface in M dwarfs and the photosphere is cooler, making it easier for the flare to have a higher contrast with the normal stellar spectrum. A recent flare on Proxima Cen brightened it by 100X in the V band! There is a rich literature regarding observations of flares on M dwarfs; it is time for a comprehensive review of it. This topic has received increased attention with the realization that there are a lot of Earth-sized planets in the habitable zones of red dwarfs, which puts them far closer to strong stellar magnetic activity, meaning many orders of magnitude more of high-energy fluxes than the Earth is subjected to (Section 7.5).

4.4 Stellar Winds

In the early 1950s Biermann noticed that comet tails always point away from the Sun and posited the existence of a solar wind. In 1958 Eugene Parker realized that structure of the magnetic field and decreasing gravity field would cause a hot thermal wind to transition from subsonic to supersonic at a few solar radii as if flowing through a Laval nozzle. This was soon thereafter observed to be the case (it is sometimes called the "Parker wind"). The structure of the corona changes with the solar cycle, being relatively more symmetric (even mostly dipolar) during solar minimum and quite structured with helmet streamers and long linear structures closer to the equator during solar maximum. Open field regions are the primary source of the solar wind that extends well beyond the planetary system until terminating against the interstellar medium. This flow of particles from the Sun is driven by the heat of the corona and so is another consequence of the solar magnetic field. The amount of mass lost by the Sun annually is negligible (about 2×10^{-14} solar masses); the Sun actually radiates away more mass through its luminosity due to fusion than it loses in the (current) solar wind.

There are two general components to the solar wind, fast and slow, which attain velocities of roughly 700–800 and 300–400 km s^{-1} respectively near the Earth (although both types can approach 500 km s^{-1}). The mix of fast and slow is a function of solar latitude, with the slow wind tending to be concentrated nearer low latitudes and during more active parts of the cycle. In some parts of the corona, most consistently near the poles during solar minimum, the field is open from near the photosphere out to several solar radii and beyond. These regions are called "coronal holes" because their X-ray emission is relatively weak due to lower temperatures near the bottom of the corona; they look dark in contrast to the hotter denser coronal closed loop regions. Examples of small ones are visible in Figure 1.1 at either limb and the active latitudes (best in the 211 Å image). The fast flows have lower densities and originate primarily in coronal holes, particularly the polar ones, and fill most of the spherical volume. The slow wind comes from the tips and edges of helmet streamers where heliospheric current sheets form and open the field outward, or from loops opening up in active regions (see Figure 4.1). The fact the Sun is rotating drags the wind structure into a trailing "Parker spiral"; at the Alfvén radius the magnetic field can no longer force the material to keep up with its photospheric roots (Figure 4.15).

The disruption of coronal loops or prominences due to photospheric motions can also cause massive ejections of solar plasma into interplanetary space (called "coronal mass ejections" or CMEs). The most common type of CME appears as a bright loop or bubble that expands with a speed of several hundred km/s. The bright feature is generally followed by a relatively dark cavity, which may in turn be followed by the remains of a disrupted prominence (Figure 4.16). CMEs are supersonic but sub-Alfvénic so a slow MHD shock forms ahead of them. The mass involved is in the range of 10^{12}–10^{13} kg, which can be several percent of the total instantaneous mass of the corona, and the energy release is comparable to a large flare. The number of CMEs directly associated with flares is relatively low, however. It is more common to see a filament/prominence become unstable and erupt. Some CMEs are not obviously associated with lower atmospheric phenomena.

Because they tend to arise from low latitudes during the active part of the cycle, they tend to be aimed in the ecliptic plane. CMEs and portions of the solar wind can

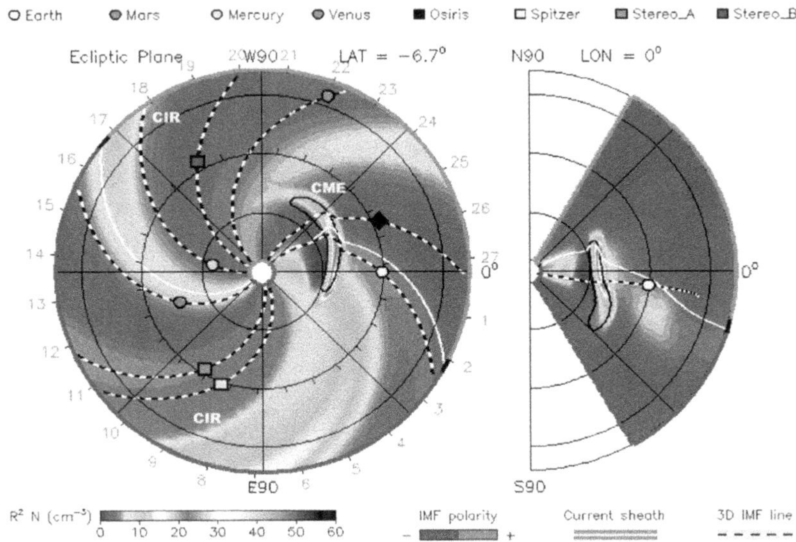

Figure 4.15. A model of the heliosphere in 2018 based on the available spacecraft measurements of the Sun and its environment. The general spiral of the wind is obvious from the polar (left) view, along with 3 denser streams and a general polarity reversal (shown in red and blue at the circumference). A CME heading toward the Earth is labeled in both top and side (right) views. Operating spacecraft (squares) that contributed observations are shown in the legend at the top, along with the inner planets (circles). Reprinted from Gibson et al. (2018), © 2018 Gibson, Vourlidas, Hassler, Rachmeler, Thompson, Newmark, Velli, Title, and McIntosh.

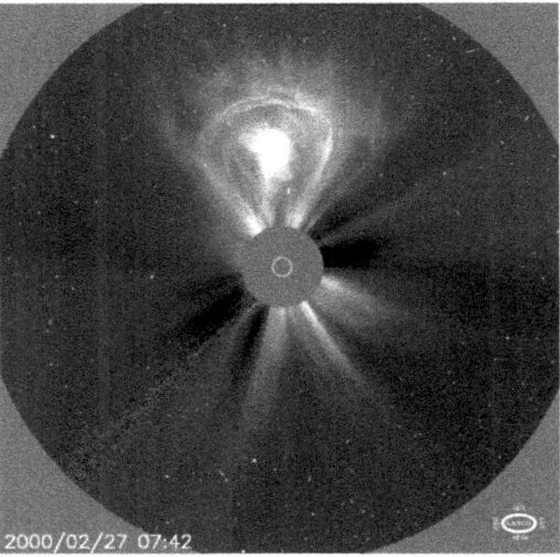

Figure 4.16. An observation of a large coronal mass ejection (CME) made by the LASCO instrument on the SOHO spacecraft. The Sun itself is represented by the white ring masked by a central occulting disk. The CME extends many solar radii and is rapidly moving outward into the solar system. Credit: Reproduced with permission from LASCO/SOHO/ESA/NASA; Joe Gurman, Simon Plunkett, Steele Hill, Stein Vidar Haugan. For an excellent video of CMEs go to https://www.esa.int/ESA_Multimedia/Videos/2020/12/Decades_of_the_Sun_as_seen_by_SOHO. Credit: MDI/SOHO/NASA.

impact all the planets (or their magnetospheres if they have one). One effect of this is aurorae on the Earth and other planets, another is possible losses from planetary atmospheres. Given that the Sun is a relatively inactive star, it is not hard to imagine that CMEs on more active stars can be far more powerful (flares certainly are). It had been suggested that perhaps much of the mass loss from M dwarfs is due to CMEs, but Wood et al. (2021) do not find convincing evidence of that. They suggest that perhaps stronger flares fail to produce CMEs because of the strong closed loops that cause the flares. We return to the effects of stellar mass loss on planets in Section 7.5.

It turns out to be very difficult to make measurements of the mass loss rates on other solar-type stars. The winds are extremely optically thin and provide no direct spectroscopic signatures. Detections of free–free radio emission from stellar winds or X-ray emission from their interaction with the ISM require much more sensitive instruments or winds that are far stronger than the Sun's. Thus detections of solar-like stellar winds have only been made using an indirect technique (which is at least two orders of magnitude more sensitive than the above methods). This makes use of the fact that the wind will eventually impact the ISM, so there is a region around the star controlled by it called the "astrosphere" (or heliosphere in the case of the Sun). A complicated interaction structure is created in the upwind direction, the direction of the star's motion toward the local interstellar medium (LISM). The Voyager spacecraft have recently encountered the termination shock of the heliosphere at about 100 au; the heliopause should be near 120 au and the outer bow shock is expected to be near 250 au.

Neutral hydrogen piles up between the heliopause and outer bow shock, aided by charge exchange between the neutral LISM and the stellar wind, so one can sometimes observe absorption from this "hydrogen wall" in the Lyα line. The amount of absorption is diagnostic of the strength of the stellar wind (given knowledge of the LISM). Such observations are complicated by absorption elsewhere in the ISM except for very nearby stars. One must also model what the stellar Lyα emission line that is being absorbed looks like at the relevant velocities. An example of such observations appeared in Figure 3.10. A little work has also been done on the Mg II resonance lines.

The interpretation of an absorption feature into a mass loss rate is quite involved and has significant uncertainty. The size of the astrosphere depends on the density of the immediately surrounding ISM and the relative velocity between it and the star. The absorption observed also depends strongly on the orientation of the line of sight between us and the star compared to the the star's relative direction into the ISM. A good review of this topic can be found in Wood (2004). The basic results show that many of the admittedly limited sample of stars with measurements have rates with a factor of a few of the solar mass loss rate, while a few active stars are one or two orders of magnitude greater (Figure 4.17). It should be pointed out that the active Sun actually loses less mass than the quiet Sun because it has more closed loops. The fact that very active stars tend to have more polar fields may play a role in how strong their winds are, but the associated angular momentum loss would be reduced by this configuration. On the other hand, there are the "slingshot prominence" cases mentioned in Section 3.3, which show the highest mass loss rates. These stars are

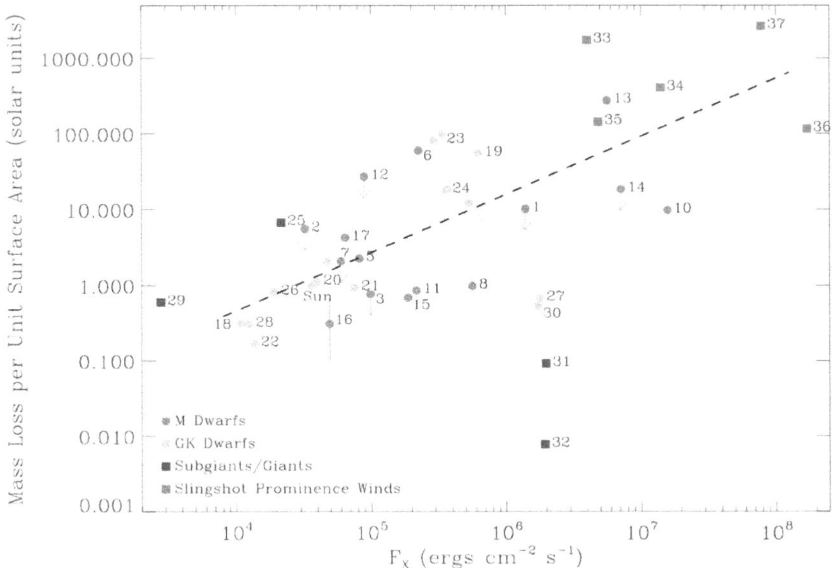

Figure 4.17. The relation between measured mass loss rates (per unit surface area) for stellar winds and their coronal surface fluxes. The "slingshot" stars are those with very rapid rotation, so that the effective gravity at the top of large prominences is much reduced. The two stars (31,32) lying well below the apparent relation (dotted line) have much larger surface areas than the others. Information on the individual stars is in Wood et al. (2021). Credit: Reproduced from Wood et al. (2021). © 2021. The American Astronomical Society. All rights reserved.

very rapid rotators and that fact helps them lose mass through prominences that lie near the co-rotation radius. Mass loss from main sequence stars is clearly an area where further work is needed and being done.

Once solar-type stars evolve off the main sequence, several fundamental changes occur. As mentioned above, there comes a point on the way to the red giant branch where stellar coronae become much weaker. Stellar winds become much stronger with mass loss rates that are 3 to 6 orders of magnitude higher while the wind velocities drop to a few tens of km s^{-1}. These wind conditions make it untenable to maintain coronal temperatures, and the winds show chromospheric temperatures instead. They become much easier to observe, for example introducing asymmetric absorption features in the Mg II resonance lines (e.g., Figure 3.6). These winds are intermediate between solar-type winds and the even more massive and slow winds observed from red supergiants (asymptotic branch stars) that are thought to be driven by radiative pressure on dust grains produced in these very luminous and low gravity stars. A general review of stellar mass loss that covers both mechanisms and diagnostics was written by Dupree (1986). It is not that easy to derive mass loss rates from spectral line profiles, generally requiring NLTE spherical radiative transfer to model the emission lines we have been talking about. Hα is one of the more commonly used lines and an example of its analysis is given by Mészáros et al.

(2009). Mass loss rates in evolved stars range in the neighborhood of 10^{-9} solar masses per year, which is five orders of magnitude greater than the solar wind.

The question of what drives red giant winds has been vexing, but there is a consensus that it very likely has to do with a different form of magnetic driving than the solar wind. Alfvén waves are the suggested mechanism. We know that red giants have magnetic fields and chromospheres, and the strong turbulence at their surfaces is likely to push energy outward via the magnetic field. Because these stars have much lower surface gravities, however, the energy does not end up confined in closed loops near the surface (as indicated by the lack of coronal X-ray emission). The field is most likely dragged outward and mostly open, and the turbulence launches waves out along the field. Alfvén waves have the right characteristics to carry the energy out some distance and then deposit it into outward momentum. The details of this mechanism, however, are not well understood since simple models fail to provide the right mass loss rates and velocity structures.

Suzuki (2007) provides one discussion of this problem. He conducts a simulation that satisfies most of the qualitative requirements. In it the wind's compressional waves are damped too quickly, but the longitudinal Alfvén waves deposit their energy in the right domain. For stars in the hybrid region he finds a structured and turbulent wind, with hot magnetic bubbles contained within the massive outflow. For stars in the red giant branch the wind launches from a lower gravity surface and the magnetic bubbles aren't formed because it is too easy to push material outward. This explains why the winds are slower and more massive for larger stars. The outstanding theoretical issues lie with the treatment of wave dissipation. It was known earlier that with the right damping length Alfvén winds work, but not clear that wave dissipation has been correctly computed (sometimes it is an ad hoc parameter). This subject will undoubtedly receive more attention in the future.

References

Ayres, T. R. 2020, ApJS, 250, 16

Ayres, T. R., Simon, T., Stern, R. A., et al. 1998, ApJ, 496, 428

Benz, A. O., & Güdel, M. 2010, ARA&A, 48, 241

Boggess, A., Carr, F. A., Evans, D. C., et al. 1978, Natur, 275, 372

Bowyer, S., Lieu, R., Lampton, M., et al. 1994, ApJS, 93, 569

Brickhouse, N. S., Dupree, A. K., Edgar, R. J., et al. 2000, ApJ, 530, 387

Chamberlin, P. C., Woods, T. N., Didkovsky, L., et al. 2018, SpWea, 16, 1470

Craig, N., Abbott, M., Finley, D., et al. 1997, ApJS, 113, 131

Davenport, J. R. A., Hawley, S. L., Hebb, L., et al. 2014, ApJ, 797, 122

Del Zanna, G., & Mason, H. E. 2018, LRSP, 15, 5

Dulk, G. A. 1985, ARA&A, 23, 169

Dupree, A. K. 1986, ARA&A, 24, 377

Engvold, O., Vial, J.-C., & Skumanich, A. 2019, The Sun as a Guide to Stellar Physics (Berlin: Springer)

France, K., Loyd, R. O. P., Youngblood, A., et al. 2016, ApJ, 820, 89

Gibson, S. E., Vourlidas, A., Hassler, D. M., et al. 2018, FrASS, 5, 32

Griffiths, N. W., & Jordan, C. 1998, ApJ, 497, 883

Güdel, M. 2002, ARA&A, 40, 217

Güdel, M., & Nazé, Y. 2009, Åpr, 17, 309

Hawley, S. L., Davenport, J. R. A., Kowalski, A. F., et al. 2014, ApJ, 797, 121

Hudson, H., & Ryan, J. 1995, ARA&A, 33, 239

Janvier, M., Aulanier, G., & Démoulin, P. 2015, SoPh, 290, 3425

Lee, K.-S., Imada, S., Watanabe, K., Bamba, Y., & Brooks, D. H. 2017, ApJ, 836, 150

Lee, L. C., & Lee, K. H. 2020, RvMPP, 4, 9

Linsky, J. L. 2018, Ap&SS, 363, 101

Liu, Y.-H., Hesse, M., Guo, F., et al. 2017, PhRvL, 118, 085101

Mészáros, S., Avrett, E. H., & Dupree, A. K. 2009, AJ, 138, 615

Moos, H. W., Cash, W. C., Cowie, L. L., et al. 2000, ApJ, 538, L1

Mount, G. H., & Rottman, G. J. 1981, JGR, 86, 9193

Ness, J. U., & Jordan, C. 2008, MNRAS, 385, 1691

Okamoto, S., Notsu, Y., Maehara, H., et al. 2021, ApJ, 906, 72

Oranje, B. J. 1986, A&A, 154, 185

Petschek, H. E. 1964, Proc. AAS-NASA Symp., The Physics of Solar Flares (Washington, DC: National Aeronautics and Space Administration, Science and Technical Information Division), 425

Pye, J. P., Rosen, S., Fyfe, D., & Schröder, A. C. 2015, A&A, 581, A28

Rosner, R., Golub, L., & Vaiana, G. S. 1985, ARA&A, 23, 413

Rosner, R., Tucker, W. H., & Vaiana, G. S. 1978, ApJ, 220, 643

Scelsi, L., Maggio, A., Peres, G., & Pallavicini, R. 2005, A&A, 432, 671

Schrijver, C. J., Lemen, J. R., & Mewe, R. 1989, ApJ, 341, 484

Stassun, K. G., van den Berg, M., & Feigelson, E. 2007, ApJ, 660, 704

Stern, R. A., Schmitt, J. H. M. M., & Kahabka, P. T. 1995, ApJ, 448, 683

Suzuki, T. K. 2007, ApJ, 659, 1592

Walter, F. M., Gibson, D. M., & Basri, G. S. 1983, ApJ, 267, 665

Wood, B. E. 2004, LRSP, 1, 2

Wood, B. E., Müller, H.-R., Redfield, S., et al. 2021, ApJ, 915, 37

An Introduction to Stellar Magnetic Activity

Gibor Basri

Chapter 5

The Evolution of Stellar Activity

5.1 Historical Introduction

The first observation of stellar (not solar) activity is probably the record on photographic plates of hydrogen emission from AT Mic that was bright on June 23 but weak on 1895 June 29, published much later by Luyten (1926). This is the first record of a flare on a young red dwarf. Objects like this are now called flare stars; the emission line results in their spectral classification of dMe and their flares are relatively bright and frequent compared with older M dwarfs. From solar studies the importance of the Ca II H&K lines had already been established by then. A few astronomers began to study the Ca II lines in stars using photographic low resolution spectra starting in the 1930s and continuing until the invention of electronic photometry.

The beginning of the serious study of stellar activity perhaps lies with the paper by Olin Wilson in 1963 (Wilson 1963) that has the title "A Probable Correlation Between Chromospheric Activity and Age in Main-Sequence Stars." Wilson looked at four young star clusters including two that have continued to play a vital role in this field: the Pleiades and the Hyades. Based on main-sequence turnoff observations, their ages were thought at the time to be 50 Myr for the Pleiades and about 500 Myr for the Hyades, Coma, and Praesepe. He also had observed about 100 field stars by then, which are expected to be generally much older (several Gyr). His basic conclusion was that younger stars have stronger calcium emission and that stellar activity must therefore decline with age. He noted that this implied the young Sun must have been much more active and subjected the planets to stronger ionizing fluxes. Finally, he correctly concluded that the underlying reason must be that surface magnetic fields themselves decline with age, although he did not know why. He then began the spectrophotometric study of Ca II emission in the time domain on 91 main sequence stars in 1966. This long-term study was the source of the "S-index" as a measure of the amount of emission and utilized in many later studies. Although initially Wilson did not detect significant variability it began to appear after a few years.

doi:10.1088/2514-3433/ac2956ch5   5-1

The next advance came when Robert Kraft published a paper in 1967 (Kraft 1967) with the title "Studies of Stellar Rotation V. The Dependence of Rotation on Age Among Solar-type Stars." Cognizant of Wilson's work he returned to the Pleiades, Hyades, and the field with higher spectral resolution (though still photographic), allowing a more sensitive measurement of Doppler broadening due to rotation. Kraft was able to draw several of the fundamental conclusions we have on stellar activity: (1) for stars with surface convection zones, more rapid rotation leads to stronger magnetic activity; (2) as solar-type stars age, magnetic winds remove angular momentum, which reduces the stellar rotation and magnetic fields; (3) more massive stars without surface convection zones remain rapidly rotating because of the lack of magnetic braking winds. He did not make the physical connection between rotation and the underlying source of magnetic fields through dynamo action, but understood that solar-like magnetic field production requires both rotation and surface convection and is strengthened as rotation speeds increase and convection zones grow deeper.

In a 1972 paper that must have a record number of citations per page, Skumanich (Skumanich 1972) returned a third time to the Pleiades, Hyades, and field (represented by the Sun) along with Ursa Major, a cluster between the other two in age. It is remarkable that a paper two pages long based on four data points has had such an influence. He reanalyzed the average Ca II strengths and concluded that stellar activity declines with age as a power law with exponent -2. This type of relation is called a rotation–activity connection, and it lies at the heart of our understanding of stellar activity. Skumanich himself realized that his result rested on shaky ground, since the age of the Hyades was already beginning to be questioned. It is now thought to be closer to 700 Myr, which would break the power law as he presented it, but the age of the Pleiades was also increased by a similar factor from 40 to 70 Myr. Later it was increased again to 120 Myr (Basri et al. 1996) along with the ages of young clusters in general but by then the relation was understood to be complicated by stellar mass. The "inverse square" or "Skumanich" law remains with us today, although it is constantly being expanded upon and refined. The rotation–activity connection is a major topic of this chapter.

The 1970s saw a large increase in the sophistication with which scientists were thinking about stellar activity, and an increase in the quality and types of data that were gathered about it. The descriptions of stellar chromospheres, transition regions, and coronae in Chapters 3 and 4 were being developed during this era; as always the Sun provided the initial inspiration earlier. Magnetic dynamos were understood to be the reason for the presence of stellar magnetic fields, and the feedback cycle between dynamos, rotation, and magnetic wind braking the reason for the rotation–activity connection. The next major advance in this area came from Noyes et al. (1984; hereafter Noyes+). By then there was a large increase in the number of Ca II measurements and data had come in from UV and X-ray observations as well. Stars with a larger range of masses and ages had been measured and it was becoming clear that the Skumanich law did not hold with complete generality.

The S-index measures the flux at the bottoms of the Ca II absorption features (where emission will lie) relative to the "continuum" flux in between the lines. Its use was understood to have the significant problem that it does not only measure chromospheric Ca II emission but also depends on the contribution of the under-lying photosphere. It is particularly problematic when applied to stars of rather different effective temperatures; the photospheres will have different contrasts relative to the chromospheres. Noyes+ adopted work that references the relevant continuum to the flux at 550 nm through calibration to a set of spectrophotometric standards with different colors. This yields a color-dependent correction to S that they called R_{HK} and is equivalent to the surface flux in Ca II divided by σT_{eff}^4. Because the flux in the core of the lines includes a photospheric contribution (that is most of the flux in the case of the Sun), they added a long discussion of how to calibrate this out (which also depends on the effective temperature) and their corrected quantity is called R'_{HK}. More recent authors have undertaken a number of possible improvements to these procedures from Noyes+.

Stars of different masses/temperatures have different convection zone depths/velocities that should affect their dynamo action along with rotation. Rotation enters through the Coriolis force, so what is important physically is the rotational velocity compared with the convective velocity. Equivalently one can compare timescales, for example using the well-known dimensionless fluid dynamical variable called the Rossby number $R_0 = P_{rot}/\tau_c$, where τ_c is a convective overturn time. Because stellar distances were not precisely known for most stars (this has changed in the Gaia era), a practice had also developed of characterizing activity levels as a fraction of the bolometric stellar luminosity rather than directly as luminosities or surface fluxes. Noyes+ noted that the spectral type dependence of the rotation–activity relation (which was not yet fully established) could be significantly reduced by a judicious set of choices in both the activity and rotation variables being considered. Essentially they added a spectral type or mass dependence to each of them by using the activity luminosity to bolometric luminosity ratio R'_{HK} and Rossby number. This set of variables has become firmly established as the best for displaying rotation–activity relations.

Before continuing down this well-trodden path, there are a couple of caveats that often are forgotten. Regarding luminosity ratios, it is not clear why the level of luminosity from a diagnostic of magnetic activity should depend in a specific way on the bolometric luminosity. The chain of physics leading from the production of magnetic fields via convection and rotation in the interior to heating by currents or waves in the upper stellar atmosphere is long and convoluted (Noyes+ shared this concern). The amount of energy so converted is in the extreme no more than 0.1% of the bolometric energy and usually orders of magnitude less. One can easily list effects along the chain that could modulate its final output. It is metaphorically like having a faucet mounted on a very large water storage tank and suggesting that the amount of water coming out of the faucet depends primarily on the size of the tank, while ignoring the fact that the faucet can be turned on by various degrees and the flow never significantly reduces the water volume in the tank. The analog of the flow through the faucet might perhaps be better represented by the average surface flux in a magnetic activity diagnostic. That expresses how much energy is emitted per unit area on the star, allowing larger stars to emit more in principle but also reducing it by the activity filling factor.

The other possible issue is with the Rossby number. Convective velocities are quite variable at different levels in the stellar interior, and the dynamo probably operates partly at the tachocline for cyclic ($\alpha\Omega$) dynamos and throughout the convection zone for turbulent (α^2) dynamos. Both are at work in solar-type stars but only the latter for fully convective stars, although we will see later that rotation still plays a role in even these stars. It is not clear that a single convective overturn time could characterize the magnetic field output, and even less so the luminosity in stellar activity diagnostics. Noyes+ elected to use the turnover time one scale height above the bottom of the convection zone, based on relatively simple (mixing length) convection theory and spent some effort on this issue. They tested the "rather ill-defined parameter that occurs in the convection zone models, and which significantly affects the calculated convective turnover time", namely the ratio of mixing length to scale height (also called α). This parameter provides a sort of "fudge factor" to adjust the relation, and later authors felt free to adjust it to make their relations as tight as possible, particularly for cooler stars that were not properly covered by the turnover time estimates provided in Noyes+ (eg. Kiraga & Stepien 2007). Reference is made to the "empirical" Rossby number to reflect this flexibility. An attempt to calculate a global theoretical Rossby number is made by Kim & Demarque (1996). Browning (2008) provides a nuanced discussion of Rossby numbers in actual stellar and dynamo models. Despite the uncertainties in how to properly calculate a Rossby number, there is a long record of it being used with luminosity ratios as a successful pair of variables that generates sensible-looking rotation–activity relations across a variety of activity diagnostics.

The essence of Noyes+ seminal paper is that R'_{HK} is broadly different in fast vs slow rotators, with a mild color (stellar mass) dependence. The relation between R'_{HK} and the log of the rotation period has a lot of scatter, although if one instead uses the surface flux (multiplying R'_{HK} by σT^4_{eff}) then the relation becomes fairly clean. However, the relation becomes even cleaner if one replaces the log of the rotation period by the log of the Rossby number, so long as one chooses the best value of α. It is expected from many forms of dynamo theory that the Rossby number should be a relevant parameter, so this result is satisfying from a physical point of view. It also removes the color-dependence of the relation. The best-fitting value for α turns out to be 2, although its best value is unity in when considering solar evolutionary models that try to match the current Sun. Noyes+ are honest about the uncertainties in what they did and the ways in which the results can be over-interpreted or misleading. Nonetheless, their basic paradigm has become "classic" and continues to be a useful way to characterize the rotation–activity connection. It has been successfully applied to diagnostics that do not suffer from the corrections needed to construct R'_{HK} (like X-rays), and so apparently has greater generality.

5.2 Rotation–Activity Relations

The pioneering studies of Ca II by Olin Wilson (coincidentally, at the Mt. Wilson Observatory) mentioned above formed the basis of big advances in studies of stellar activity. He used a spectrometer to feed an electronic photometer in three narrow bandpasses; one centered on the Ca II K line and two on the "continuum" regions to either side of the line (Figure 3.3). Although a somewhat imprecise measure of the purely chromospheric contribution in the K line, the "S-index" he defined was more precise and repeatable than any previous observations. More importantly, Wilson undertook a long-term monitoring program on nearly 100 stars. This would enable him to see how variable they are over time and he hoped to detect stellar rotation and activity cycles. Eventually he detected signs of cycles in about a quarter of his sample. A summary of his program and its continuation at Mt. Wilson can be found in Baliunas & Vaughan (1985). That was the source of the data for Noyes+. They summarize what had been learned to that point about stellar cycles and calcium variability.

Another long-term program of monitoring both broadband colors and Ca II emission in over 70 stars was begun in 1992 at the Lowell Observatory. Results from it are summarized by Radick et al. (2018). In addition, there is a rapidly increasing set of high resolution spectra from which S-indices can be derived being collected by echelle spectrometers used to hunt for and study exoplanets and their stars. In particular, the Keck and HARPS surveys have observed thousands of stars at least a few times and with good long-term monitoring in some cases, including Wright et al. (2004) and Boro Saikia et al. (2018; which includes references to several large previous data sets). For cooler K and M stars, surveys using Hα as the activity diagnostic are better. Large such surveys include Gizis et al. (2002), West et al. (2008), and Jeffers et al. (2018).

Since the appearance of Noyes+, there have been a steady stream of papers revisiting the rotation–activity connection with better and larger samples of stars, and in the Gaia era much better sets of stellar parameters. Later versions of the S-index made with different instruments and its calibrated conversion to R'_{HK} are the subject of continuing work. I only discuss a few of the latest ones here; the history of the subject can be followed in reference chains back from them. A good discussion of the continuing quest for the right form of both the Rossby number and the Ca II activity diagnostic can be found in Mittag et al. (2018). That paper goes into the additional question of whether there is a non-magnetic (basal) component of the chromosphere that should be subtracted off along with any photospheric radiation when computing the Ca II emission that is due to magnetic fields. The possibility of a basal chromosphere has always been on the table, since acoustic waves generated by convection will steepen and dissipate in shocks given the steep density gradient outward. In the 1960s there were serious suggestions that acoustic input could be a major ingredient in chromospheres though that is no longer thought to be the case. It is certainly clear on the Sun that Ca II emission and magnetic fields are very closely related.

The summary in Mittag et al. (2018) of the various determinations of empirical Rossby numbers is that the Noyes+ relation does a reasonable job of providing a functional form that works, although there is a need to multiply it by a factor of about 2.5 to provide the best fits for a variety of stars and diagnostics. The basic relation between rotation and activity seen using the Rossby number is that there is a power law decline in activity starting somewhere around $\log(R_0)$ of -1 toward more positive values, and the situation for $\log(R_0)$ greater than zero becomes less and less clear because of the difficulty in determining the rotation periods and low activity levels of stars with $\log(R_0)$ of about unity or larger. Although there are catalogs of thousands of Ca II activity measurements (e.g., Boro Saikia et al. 2018), the number of those stars for which a rotation period has also been measured is much smaller, and oddly deficient in stars with $\log(R_0)$ less than -1.

Mittag et al. (2018) also address the question first posed by Vaughan & Preston (1980) as to whether there is really a gap (named after them) in the distribution of active and inactive stars that they saw in their early small sample. The presence of such a gap could be due either to a recent burst of star formation (which they rejected) or a relatively rapid transit of stars from active fast rotators to significantly less-active slow rotators. The presence of the Vaughan–Preston gap has been debated since then, with some papers finding further evidence for it and others finding that it is not real but just due to particular sample biases. With their very large sample Mittag et al. (2018) show that stars with B–V colors less than about 1.1 tend to be dominated by $\log(R'_{HK})$ values lower than about -4.75 while stars redder (and lower in mass) are mostly higher than that. Red dwarfs with B–V greater than about 1.4 are found roughly equally on both sides of the boundary. The composition of individual smaller samples could therefore show the gap or not depending on their composition.

Lehtinen et al. (2020) provide newly compiled observations of more than 200 stars (still from the Mt. Wilson project) that encompass both main sequence and evolved field stars. Their modern version of the Noyes+ Ca II relation can be seen in the top panel of Figure 5.1. The stellar parameters are much more precise than accessible to Noyes+ because they come from Gaia. They particularly wanted to address the question of whether use of the Rossby number can unify the relation even though convection zones change as stars evolve off the main sequence, and whether relations that only depend on direct stellar parameters cannot bridge the gap. They therefore use Rossby numbers based on theoretical stellar interior calculations. The original Noyes+ plot covered a range in $\log(R_0)$ from -0.6 to 0.4 and fit a quadratic to their (50 main sequence and subgiant) points. It is apparent in the top panel that in that part of the diagram a simple power law does reasonably well (albeit with a scatter of nearly half a dex). It is also apparent that outside that range of $\log(R_0)$ the relation is more uncertain and not consistent with that power law or a quadratic.

It is helpful to look at activity diagnostics that are not affected by photospheric corrections and that are cleaner measurements than Ca II emission. The Mg II lines would be superior to Ca II if there were enough observations of stars with rotation periods. X-rays are much better in this context and more sensitive to changes in activity. There are now enough observations to make them quite useful for this purpose and this situation should get even better with eROSITA. For stars with $\log(R_0)$ less than -1, $\log(L_X/L_{bol})$ tends to become "saturated," meaning it retains a relatively constant value independent of Rossby number that is much higher than the Sun's. This has been demonstrated by Pizzolato et al. (2003), Kiraga & Stepien (2007), Reiners et al. (2014), and Wright et al. (2018) among others. Activity expressed as $\log(L_X/L_{bol})$ saturates at a level around -3, or about 1000 times greater

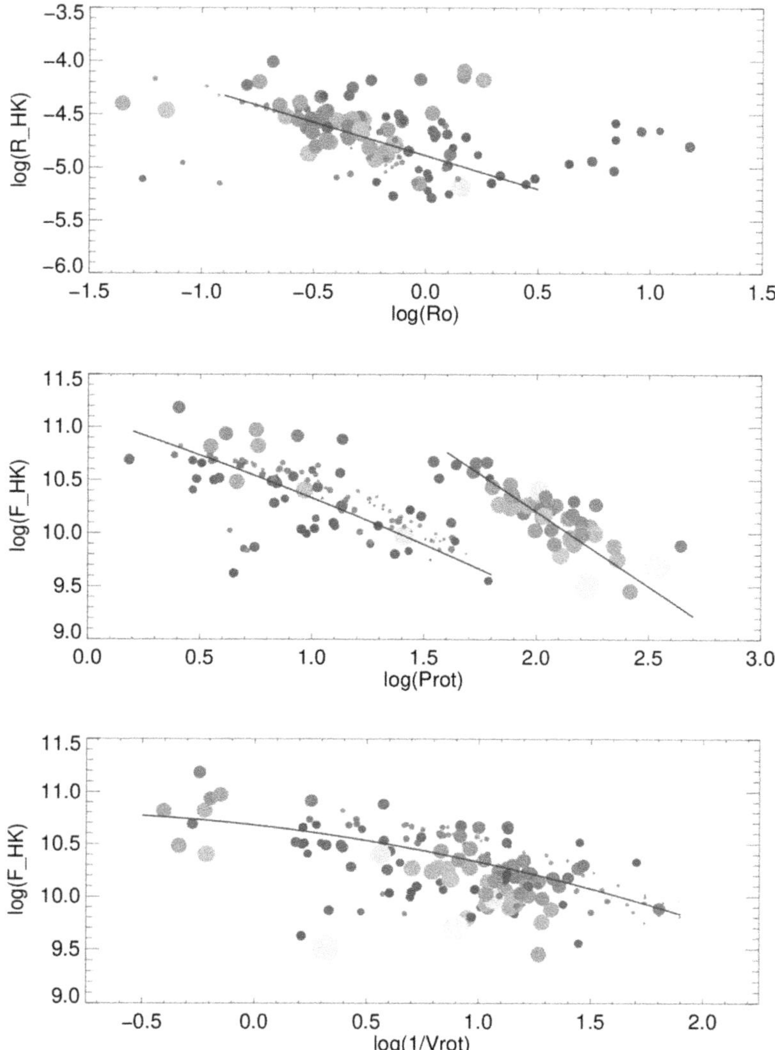

Figure 5.1. Three versions of Ca II rotation–activity relations, adapted for this book from the data presented in Lehtinen et al. (2020). The logarithmic span of each is the same. The top one uses the classic Noyes+ variables, R_{HK} and R_0, the second uses surface flux vs rotation period, and the third replaces the period with the inverse surface rotation velocity. Both the colors and sizes of the symbols are driven by stellar luminosities (more luminous are larger).

than the Sun, for $\log(R_0)$ less than about -1. For higher Rossby numbers the activity decreases roughly like the inverse square of the Rossby number, with a slope of -2 in the log–log plane as can be seen in Figure 5.2. Hα in M dwarfs (after adjustment for its behavior in inactive stars) also shows saturation/linear drop behavior similar to X-rays (Newton et al. 2017).

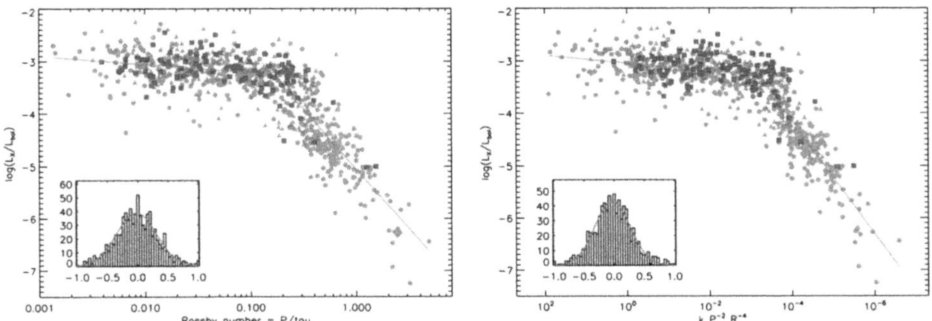

Figure 5.2. The relation between $\log(L_X/L_{\rm bol})$ and two combinations of variables including rotation and stellar parameters. The left panel uses Rossby number (period and convective overturn time), while the right panel uses period and stellar radius. The colors represent stellar ages—blue squares: very young stars (up to 50 Myr); green triangles: young stars (between 85 and 150 Myr); magenta triangles: intermediate-age stars (600–700 Myr); red circles: field stars. The histograms show the residual errors from the simple power law fits. Credit: Reproduced from Reiners et al. (2014). © 2014. The American Astronomical Society. All rights reserved.

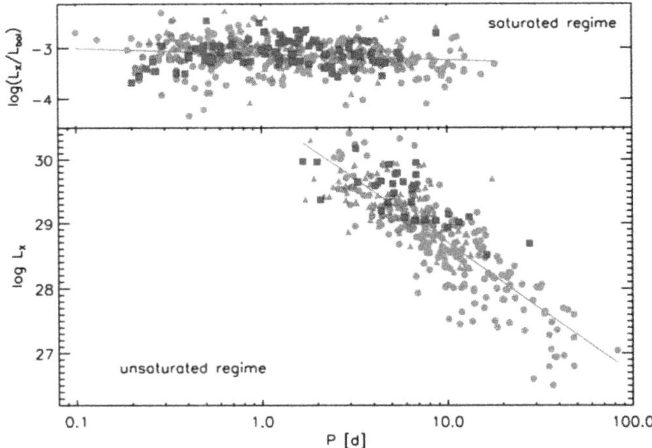

Figure 5.3. Another way of depicting a relation between coronal activity and rotation period. The same stars and colors are repeated from the last figure. In this case we see a good direct relation between X-ray luminosity L_X and rotation period for unsaturated stars, while the saturated stars all show the same value of $\log(L_X/L_{\rm bol})$ without regard to period. The upper relation does not work for L_X (so there is a mass dependence). Credit: Reproduced from Reiners et al. (2014). © 2014. The American Astronomical Society. All rights reserved.

The question of whether the use of luminosity ratios and Rossby numbers produces tighter rotation–activity relations than surface fluxes or luminosities combined with more straightforward metrics of rotation has been addressed by two newer papers. Reiners et al. (2014) re-examined the X-ray relations. In addition to the traditional Noyes+ variables shown in the left panel of Figure 5.2; they compare the X-ray luminosity (not normalized by the bolometric luminosity) against a combination of period and radius power laws with exponents that minimize the scatter in the relation. The combination of $P^{-2}R^{-4}$ leads to a fit that is nearly the same but slightly better than the Noyes+ variables. They explain this as perhaps due to the fact that convective overturn times scale something like $L_{\mathrm{bol}}^{-1/2}$ and $T_{\mathrm{eff}} \propto R^{1/2}$ approximately, so one can imagine rearranging variables so that the two relations are roughly equivalent. Of course, surface flux and luminosity are also related through R^2. Newton et al. (2017) make a similar test for Hα in M dwarfs; they point out a mass dependence for unsaturated stars when using $P^{-2}R^{-4}$ in that case.

In Figure 5.1 from Lehtinen et al. (2020) there is another attempt to see whether one combination of variables is more fundamental than another. The top plot is the canonical Noyes+ version while the middle plot uses surface flux on the ordinate. Noyes+ noted that using axes of $\log(F_{\mathrm{HK}})$ and $\log(P_{\mathrm{rot}})$ was nearly as tight for their sample. The middle plot shows that is true for main sequence stars, but giants follow an offset tight relation in the sense that a longer period produces the same Ca II emission. In the top panel the two classes of stars mix together when using luminosity ratio and Rossby number, which is a good argument for using that relation. The difference between them is related to the different influences of stellar radius and temperature, which are related to bolometric luminosity and convective overturn times.

In the bottom panel of Figure 5.1, however, I used the same data as in the top two panels to show that the rotation–activity relation can be re-unified for all stellar luminosities when using the surface flux on the ordinate and simply dividing the rotation period by the stellar radius (producing the inverse surface rotation velocity) on the abscissa. The bottom plot is a bit superior to the top plot; in the top plot the stars above $\log(R_0)$ of 0.5 don't really fit and it contains 50 fewer stars because the authors didn't trust some of their Rossby numbers. This is something of a reprise of my comments at the Cool Stars Workshop in 1985 titled "Rossby or Not Rossby?." It will be interesting to look at the X-ray relations with eROSITA data, Gaia stellar parameters, and these same variables.

The question of which is the fundamental set of variables may or may not be a fruitful line of investigation to pursue further. It is clear that rotation and activity are related to each other in a way that is mediated by stellar parameters. The empirical Rossby number has a conceptual justification related to dynamo action (which is certainly relevant). The Noyes+ relation compares a luminosity ratio to a temporal ratio. Alternatively, the surface flux may be more directly related to a surface rotation variable and yields a similar-looking relationship between rotation and activity.

As one moves to lower mass stars, Ca II becomes increasingly difficult to observe as the brightness in the violet end of the visible spectrum rapidly decreases. Although Hα has ambiguities in interpretation due to its NLTE formation (Section 3.3), it is a popular activity diagnostic for low mass stars. For them it tends to yield results consistent with the other activity diagnostics (Newton et al. 2017; Schöfer et al. 2019), and shows a similar decrease with slower rotation. The stars without emission have periods preferentially longer than about 10 days. The presence of stronger Hα absorption in early M dwarfs can indicate more activity than weaker absorption

(Section 3.3), but as activity levels increase the line goes into emission, and does so more easily as one looks at cooler stars.

The fraction of stars showing emission is about a quarter for early M dwarfs, about a half near M4, and rises above 0.8 at M8 and later (Jeffers et al. 2018). Of note is the lack of any discernible feature in Figure 5.4 at the point where stars

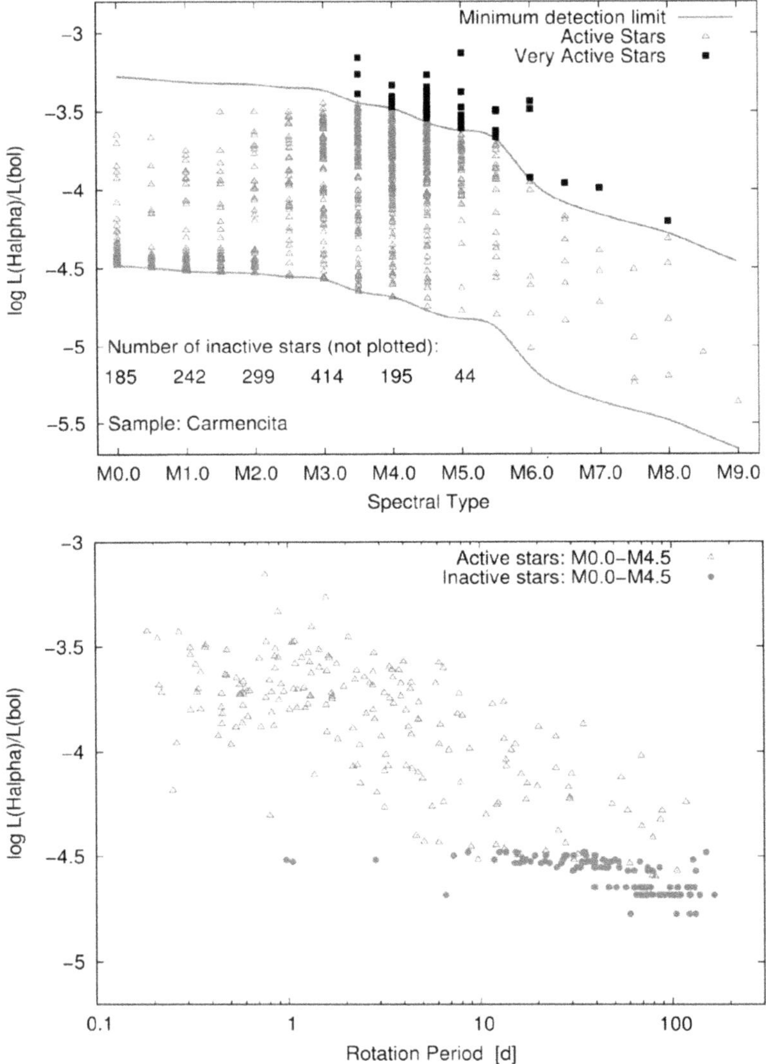

Figure 5.4. The behavior of Hα in M stars from the augmented CARMENES spectroscopic survey. The top panel shows the number of stars with Hα emission in each spectral sub-class; the numbers of stars in each bin not showing emission are indicated below the red line (none at M6 and later). The normalized emission peaks in the mid-M range. The bottom panel shows the dependence of normalized emission on rotation (inactive stars in red). Credit: Jeffers et al. (2018), reproduced with permission © ESO.

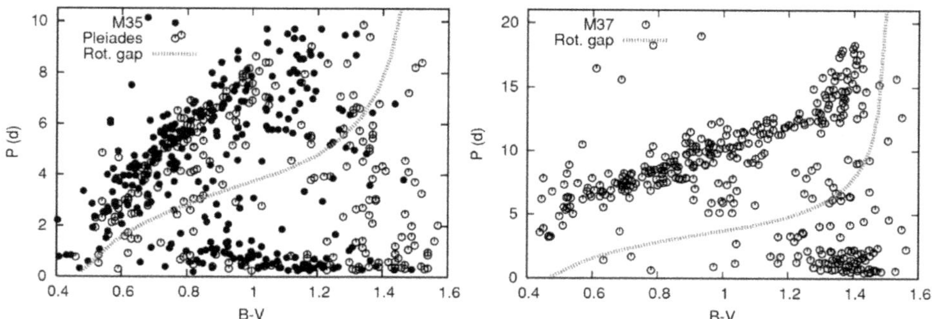

Figure 5.5. The proposed gyrochronology relation. The left panel has data from the Pleiades (120 Myr; open points) and M35 (150 Myr; solid points). The right panel has data from M37 (550 Myr). The I sequences are above the "rotational gap" (indicated in color) and the C sequences are below it. Note the difference in the two period scales. Credit: Reproduced from Barnes (2010). © 2010. The American Astronomical Society. All rights reserved.

become fully convective (M3 and later). These authors show that normalized Hα luminosities are fairly similar to normalized X-ray luminosities, exhibiting saturation for Rossby numbers less than 0.01 and perhaps even supersaturation (decrease toward smaller Rossby numbers) for the most active stars. The greatest activity levels are seen for mid-M dwarfs. The emission fraction decreases rapidly as one moves into L dwarfs for a different reason; this topic is revisited in Section 7.1.

5.3 Age–Activity Relations and Gyrochronology

Magnetic activity is related to stellar rotation but also causes the star to lose angular momentum over time via braking from a magnetic wind. This suggests that there must be rotation–activity–age relations that probably also depend on stellar mass. That stellar rotation is influenced by magnetic activity is clear from the difference between the rapid rotation on the main sequence of hot stars (those without surface convection zones) and the much more slowly rotating convective main sequence stars. But as already described above, there is a relation between the rotation and activity of cool stars and older cool stars are less active. This offers the promise that one might be able to constrain stellar ages based on either their activity levels or their rotation periods. It is already apparent that these relations will also depend on stellar mass, given the need to include basic stellar parameters in the rotation–activity relations.

Stellar rotation is a more straightforward parameter to measure than stellar activity. It can be measured in a single observation if rotational Doppler broadening is apparent in a spectrum of sufficient resolution. This generally requires a spectral resolution of at least 20000 (defined as the ratio of a wavelength to the width of the spectrometer point-spread-function at that wavelength). Doppler broadening was the primary means of gathering stellar rotations in the 20th century. Unfortunately it does not actually measure the rotation period but rather the projected rotation velocity $v \sin(i)$. To translate this into a rotation period requires knowledge of both the stellar radius and the inclination of the rotational pole to the observer; the latter is generally unknown. Many early studies of rotation–activity–age relations were therefore plagued by having only lower limits to the rotation periods. Monitoring actual activity indices (like Ca II) can yield a period but that was restricted by the difficulty of gathering enough observations with a spectrometer. This had to be done one star at a time until the advent of multi-object spectrometers. As photometric monitoring by robotic telescopes became possible the number of directly observed periods (due to starspot modulation) increased substantially, and modern space-based photometers have greatly increased our catalog of rotation periods.

The other problem in calibrating a rotation–activity–age relation is that one needs an independent way to determine stellar ages in order to calibrate the relation. The main method has been to utilize stars in open clusters since their ages can be determined by upper main-sequence turnoffs, or by lithium dating (up to about 200 Myr). It is also possible to make a determination (with more uncertainty) based on the isochrone position in stellar evolution models of solar-mass field stars if you know their luminosity (brightness and distance) and metallicity. Recently it has additionally become possible to get a (less uncertain) evolutionary age from asteroseismology for a few hundred main sequence stars (Chaplin et al. 2014) and this sample will keep getting larger. Evolutionary methods do not work for stars less than about 0.8 solar masses, however, because their evolutionary timescales are increasingly longer than the current age of the universe. These same stars are difficult asteroseismology targets because of their faster frequencies and low-amplitude integrated pulsations.

Instead, one can utilize Skumanich-type relations between age (τ) and activity; recall that he proposed that activity decreases like the square of the age. These are calibrated against stars in clusters but can then be applied to field stars, yielding a "chromospheric age." By the early 2000s enough actual rotation periods were known for stars in young clusters along with the periods from the Mt. Wilson sample. Barnes (2003) showed that there is a pile-up of stars that have had enough time to brake significantly along a locus in a plot of the Skumanich law represented by $\log(P_{\text{rot}}/\tau^2)$ vs stellar mass (using color as a proxy for mass) that he called the "I" (interface) sequence. Although he chose this name in reference to "interface dynamos" you could think of it as the Inactive sequence, since there is a second set of stars off the relation that have shorter periods and are more active. In this paper he also suggested calling rotation–age relations "gyrochronology," which is now the accepted terminology.

In an extension of that work, Barnes (2010) incorporated the more active stars, which have a more complicated relation between rotation, activity, and age. He labeled these stars as belonging to the "C" (convective) sequence, given the idea that their dynamos might operate primarily throughout their outer convective zones instead of at the interface between their radiative and convective zones called the tachocline. This topic is discussed in more detail in Section 6.1, revealing the situation to be more complex than implied by these two names. The Sun certainly has both types of dynamo in operation and their relative dominance changes with the activity cycle. I suggest thinking of the C sequence as simply referring to younger and more active stars. There is a tendency for these stars to have predominantly lower mass (larger B–V).

Regardless of labels, Barnes recognized that empirically the behavior of the C sequence requires a different equation than for the I sequence. He incorporates two specific models for the period decrease through magnetic braking with as similar a form as possible. These explain the difference between the two sequences by using the Rossby number for the I sequence and inverse Rossby number for the C sequence, in the following relation for period change:

$$\frac{dP}{dt} = \left[\frac{k_{\rm I}P}{\tau} + \frac{\tau}{k_{\rm C}P} \right]^{-1}. \tag{5.1}$$

The constants, derived empirically, are $k_{\rm C} = 0.646$ days Myr^{-1} and $k_{\rm I} = 452$ Myr days^{-1}, and τ is the age in Myr. The behavior of these functions means that stars move fastest near the C/I boundary, which generates a "rotational gap" that has some resemblance to the Vaughan–Preston gap. This is in contrast to the results mentioned earlier but from a much more heterogeneous sample by Mittag et al. (2018). Another attack on the same problem preceding the later Barnes paper above was performed by Mamajek & Hillenbrand (2008). They used more clusters than Barnes and re-calibrated the precursor gyrochronology relation by him. They altered the values of the constants in his relations somewhat to obtain better fits to their data, but the underlying approach is the same. A good collection of references on the existence and causes of the two branches can be found in the introductory material in Sood et al. (2016).

Mamajek & Hillenbrand (2008) also address the question of chromospheric ages. They re-calibrated the log(R$'_{\rm HK}$) versus log(τ) relation as a function of color by using cluster data to define an activity–color relation for each one and produce a value of log(R$'_{\rm HK}$) at the color of the Sun in each case. A polynomial was then fit to each of the relations predicting one variable from the other. This exercise suffered from the same paucity of clusters older than 1 Gyr and adds the Sun as an older point as previous studies had. In addition, they added a number of binaries containing solar-type stars whose ages can be estimated from isochrones. They re-examined the correlations between Rossby number and activity measured by Ca II and X-rays. For stars in the I sequence and for unsaturated coronal stars, they find nice tight fits. These can then be translated to relations between age and activity that relate activity, Rossby number, and age as a function of color (stellar mass) but they are only valid between B–V colors of 0.5–1.0. The one cluster with an age older than 1 Gyr, M67 at about 4 Gyr, does not fit as well onto their relations. They note that they had fewer rotation periods for M67 and a number of its members are hotter than the Sun, where their relations were not as well-defined.

The understanding of gyrochronology began receiving a large dose of help soon afterward with the launch of the Kepler mission. It supplied more than 30,000 new photometric periods for stars on or near the main sequence, and was also able to observe with much better detail stars in a few older clusters. This was augmented by further data from the K2 mission that included M67 and other clusters. Not only were the ideas behind gyrochronology nicely confirmed, but the sample size, precision, and age coverage was greatly improved. In particular this supplied a lot more detail on how stars of different masses spin down in clusters of different ages, and the sample was extended well into the M dwarfs (there was rather little information on them previously). Figures 5.6 and 5.7 summarize the new data as of 2020.

Included on Figure 5.6 are dashed theoretical curves from Spada & Lanzafame (2020) who calculate angular momentum losses from stars of different masses as they evolve through these ages. They include magnetic wind braking along with the

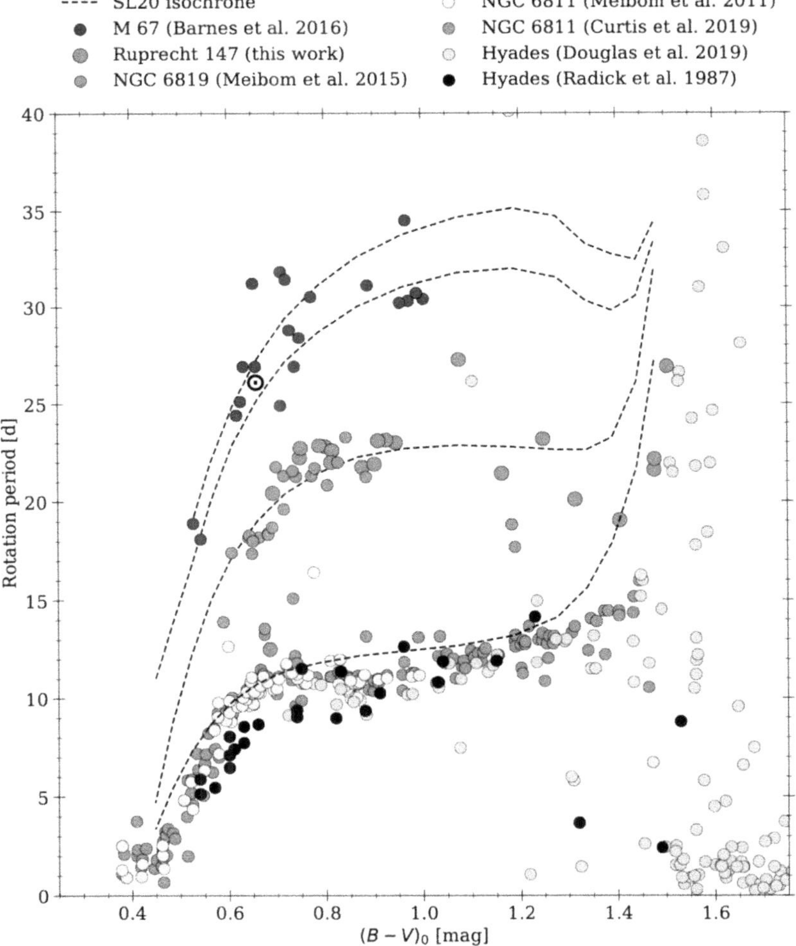

Figure 5.6. The rotational behavior of stars of different masses (represented by the B–V color) in older clusters of different ages as gathered by the Kepler mission. The ages of the clusters are M67 (4.0 Gyr), NGC 6819 & Ru147 (both 2.5 Gyr), NGC 6811 (1.0 Gyr), and the Hyades (600 Myr). The Sun is also shown (4.5 Gyr). The dashed lines are isochrones from Spada & Lanzafame (2020) that represent slowing from ages of 1, 2.5, 4.0, and 4.57 Gyr. Credit: Gruner & Barnes (2020), reproduced with permission © ESO.

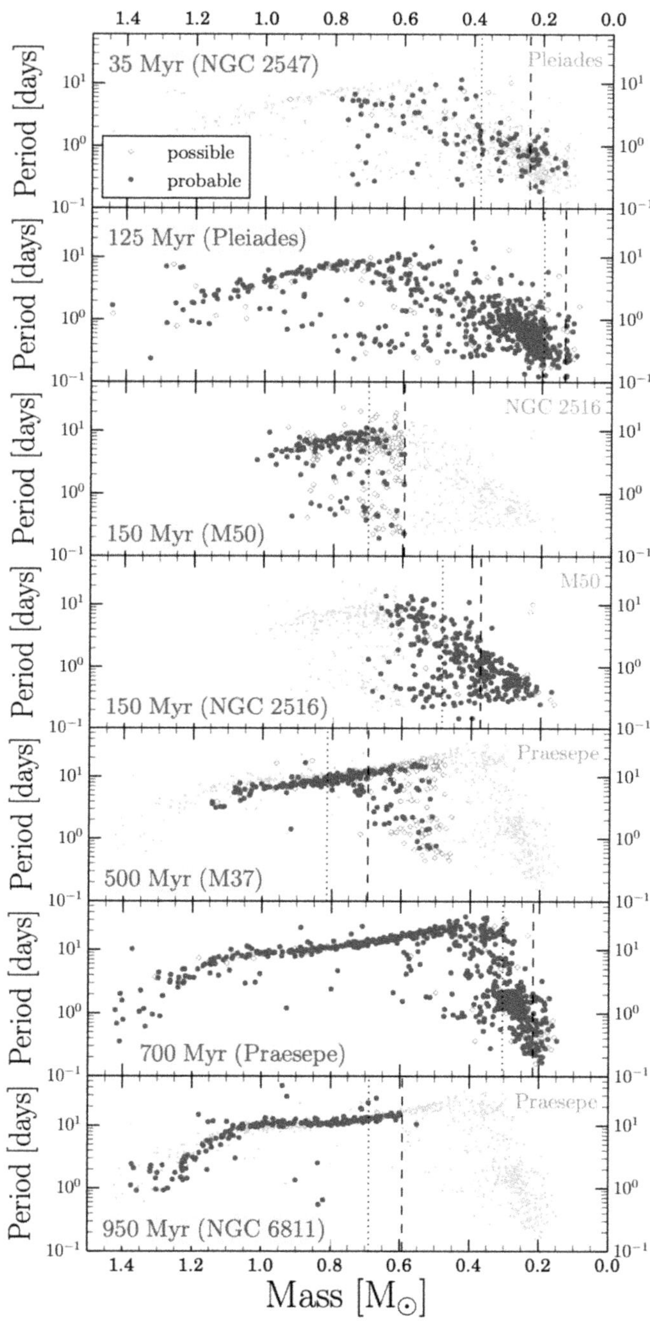

Figure 5.7. The rotational behavior of clusters of different ages from several ground-based surveys and the Kepler spacecraft. Membership and stellar parameters have been carefully vetted using Gaia. Points are probable members (blue), possible members (green), and comparison clusters (gray). The vertical dashed and dotted lines are observational magnitude limits. Credit: Reprinted from Godoy-Rivera et al. (2021).

addition of angular momentum as the more rapidly rotating core couples to the evolving convective envelope (Bouvier et al. 2014). Work over the past 20 years on this question has proceeded on several fronts, particularly in regard to the behavior of young stars starting with observed distributions of rotation periods in star-forming regions. Another modern front has opened on the question of how the magnetic field configuration evolves over time. As discussed in Section 4.2 stellar winds originate mostly from open field regions, so if the fraction of open versus closed fields changes over time the coupling of the field to angular momentum losses will also change. A discussion of this can be found in Finley & Matt (2017); their primary conclusion is that the most important parameter is what fraction of the total magnetic field is in the dipole component.

Further improved data has been gathered by several ground-based surveys, and is now being collected by the TESS mission and other surveys. At the time of writing, the best compendium of cluster rotational data has been compiled by Godoy-Rivera et al. (2021). It is always an essential part of such studies to identify actual cluster members and assess the likelihood of potential members, as well as doing as good a job as possible on their stellar parameters. The Gaia mission has made this task substantially more accurate. The authors do a careful job of these tasks, and present rotational data for stars including lower masses than before, down to 0.2 solar masses in some cases. The youngest cluster is closest to representing an initial distribution of stellar rotations. Figure 5.7 represents the state-of-the-art that can be presented in this book.

Godoy-Rivera et al. (2021) draw several conclusions. One is that stars greater than 0.5 solar masses behave in a monotonic fashion, consistent with a Skumanich-type law. There is some overlap in the periods of 0.8–1.0 solar-mass stars between 0.75–1.5 Gyr that they speculate might have to do with metallicity differences in Praesepe, NGC 6811 and NGC 752. There is increasing evidence both observational and theoretical that metallicity affects the manifestation of stellar activity (e.g., Karoff et al. 2018). Higher metallicity produces larger facular contrasts and deepens convection zones, for example. Alternatively the overlap may have to do with the core–envelope angular momentum transfer discussed above; that could cause a temporary stalling of spindown. They do not find support for the suggestion by Barnes that there is a rotational gap due to the difference between $\alpha\Omega$ (interface or tachocline) dynamos and α^2 (convective) dynamos, in the sense there is no notable feature at the mass below which stars are always fully convective (0.3 solar masses). This is not surprising since it is likely that the mix of the two dynamos changes smoothly in favor of the latter as one moves toward this boundary.

Stars with masses a little greater than 1.2 solar show greater scatter and shorter periods due to the thinning of the surface convection zone which weakens the dynamo and presages the rapid rotation of even more massive stars. There is a general tendency for stars less massive to converge to a narrow band of periods with time, and this convergence works its way from more to less massive stars with time. There is generally a point in mass at each age below which the distribution in period stops being tight at longer periods and becomes scattered down to shorter periods. Godoy-Rivera et al. (2021) suggest this means that the braking torque is weaker and less consistent in rapid rotators, which could be due to their more complex and closed field configurations (a smaller fraction of the total field in the dipole component). Another possibility is that the Rossby number might also depend on rotation at a given mass, since the spindown laws can be expressed in terms of it. At any rate, stars eventually erase the memory of their initial angular momentum and all converge to the same rotation period at the same mass and ages of a few Gyr. This convergence takes longer for lower mass stars.

Our rapidly increasing understanding of gyrochronology is still relatively weak at the time of writing for solar-type stars older than the Sun because of their slow rotation. Their rotational Doppler broadening becomes smaller than the broadening due to atmospheric turbulence. The amplitude of activity diagnostics makes them increasingly difficult to see (depending on the diagnostic) as the dynamo production of fields decreases. The detection of photometric rotation periods longer than the Sun's is increasingly difficult due to the low-amplitude variability, the increasingly aperiodic signal (Section 2.2), and the requirement for longer coverage. The Sun itself would not have yielded its period if observed by Kepler. Another issue in that case is the difficulty of detecting periods longer than a month due to their suppression by the Kepler reduction pipeline and the independent calibration of each quarter.

The large sample of Kepler rotation periods of field stars presented by McQuillan et al. (2014) shows a conspicuous absence of rotation periods longer than 35 days for stars with larger than 0.5 solar masses (Figure 5.8). They have twice as many stars with period non-detections as with detections. As mentioned in Section 2.2, the Sun is young compared to the Galaxy, so one might expect that a field sample will be composed in good part of older stars. This question has been addressed in detail by van Saders et al. (2019) who include Galactic models and careful consideration of the Kepler selection biases and reduction issues. They show that if the detectability depends on Rossby number then the threshold above which detections appear to

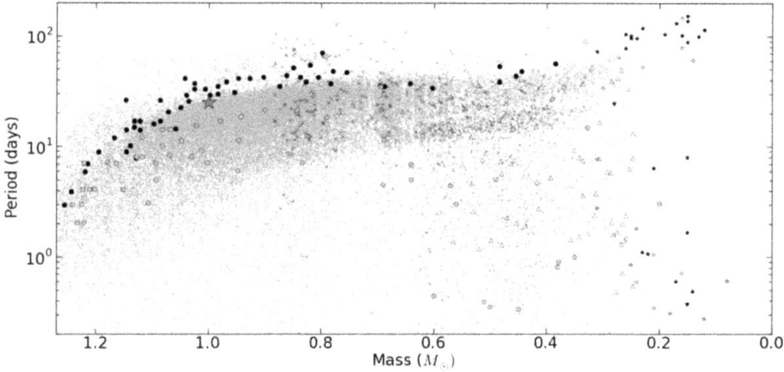

Figure 5.8. The rotational periods of a large sample of field stars as found from Kepler light curves. Additionally there are points from various ground-based surveys; kinematically older stars are black and younger ones are gray (the Sun is red). Notice the bifurcation in periods below about 0.7 solar masses. Reproduced from McQuillan et al. (2014). © 2014. The American Astronomical Society. All rights reserved.

fade is about $R_0 \sim 2$. This produces the expected effect of making the Kepler sample of detected periods deficient in older stars. They also discuss the effect of contamination by subgiants; massive (more rapidly rotating) subgiants can be mistaken for young dwarfs and slowly rotating subgiants for old dwarfs unless one has excellent stellar parameters.

Against the explanation that the long-period edge is simply a detection threshold is the fact that the amplitudes of photometric variability for stars near the edge are not necessarily the lowest amplitudes. There are stars with long periods and higher amplitudes, as noted by Reinhold et al. (2020) and mentioned earlier. There are also stars with shorter periods that have lower amplitudes. Recall from Section 2.2 that photometric amplitude may not reflect total coverage so much as spot emergence asymmetries; the latter may not depend in the same way on Rossby number. Detectability also depends on sufficient periodicity in the light curve; that is affected by spot lifetimes which could also have a different Rossby number dependence.

The other feature apparent in Figure 5.8 is a small gap that appears below about 0.7 solar masses at rotation periods between 10 and 20 days and appears to become more pronounced at lower masses. A feature like that could be produced if stars brake rapidly at those masses when they have slowed to about 10 days from faster periods. This could happen if there were some sort of dynamo phase change at those Rossby numbers, for example. An alternate explanation is that there was a relatively recent (less than 0.5 Gyr) burst of star formation producing an excess of lower mass stars that have not yet slowed since they take longer to slow down. This issue is discussed by Gordon et al. (2021) who provide the references and arguments to date. They analyzed nearly 10,000 new periods of main sequence stars from the K2 mission. It covered more than ten different pointings around the ecliptic as opposed to the single Kepler field in MAM14. They see the same gap with even greater clarity and so argue that it is not likely to be due to a burst in star formation seen in a particular direction. It must instead reflect a speed-up of the spindown for lower mass stars in this Rossby number range. There have been several other suggestions using different diagnostics that something may happen to the dynamo at these masses and Rossby number that should be pursued.

It is easier to assess the activity levels of older stars than their rotations because the activity diagnostics are more easily measured, albeit at lower values. This was discussed by Mamajek & Hillenbrand (2008) and is addressed again in this context by Metcalfe & Egeland (2019). Ages estimated from gyrochronology begin to deviate (becoming too young) compared with ages estimated from activity levels for stars significantly older than the Sun. This trend is reinforced by cases where an isochrone or asteroseismic age is available; they disagree in the same sense with the gyro-age. Following the proposition from van Saders et al. (2016), Metcalfe & Egeland (2019) agree that there is increasing evidence that the dynamo may become too weak or shift into a form that further weakens magnetic braking at something like the age of the Sun. They suggest this is because the Coriolis force becomes too weak compared with convective velocities to produce the large-scale $\alpha\Omega$ dynamo effectively. The spindown would be stalled thereafter, leaving stars rotating too rapidly compared with predictions from gyrochronology derived from younger stars.

Magnetic activity, however, appears to continue to decrease. This means that chromospheric ages could be more reliable than gyro-ages for older main sequence solar-type stars when other methods aren't better. It is not clear, however, why the magnetic field that produces this activity should keep decreasing if produced in an increasingly dominant α^2 (convective) dynamo, since the stellar luminosity (and hence convective velocities) and the depth of the convection zone both increase with age. Although evidence is presented in Section 6.2.1 that convective dynamos still care about rotation, if the spindown is stalled then that is no longer decreasing with age. The question of late stalling is also carefully discussed by van Saders et al. (2019). They agree that the long-period edge for solar-type stars could be produced in this fashion, but point out a number of caveats that mean it is not yet obviously the right explanation. This topic is clearly one that will be actively researched in the next few years.

We now turn to lower mass stars. A very recent discussion of this topic can be found in Popinchalk et al. (2021). Early M dwarfs have about half the mass of the Sun. They take longer to join the slow branch of solar-type stars, somewhere between 200 and 500 Myr (this needs further intermediate-age data for better precision), and have converged onto the sequence by 700 Myr. Mid-M dwarfs show some signs of slowing down around 1 Gyr but the convergence times for them and the late M dwarfs are not currently known. This is partly because it is hard to assign quantitative ages to field M dwarfs and hard to see them in the more distant clusters that are old enough. It is clear that braking continues to occur because the field population does not show rapid rotators among kinematically old stars. The bimodality apparent at the low mass end of Figure 5.8 could suggest that these stars move rather quickly from the fast to slow branch when they do so. Newton et al. (2017) present the active/inactive state of a general sample of field M dwarfs as a function of their rotation period. Figure 5.9 shows that the inactive stars are the slow

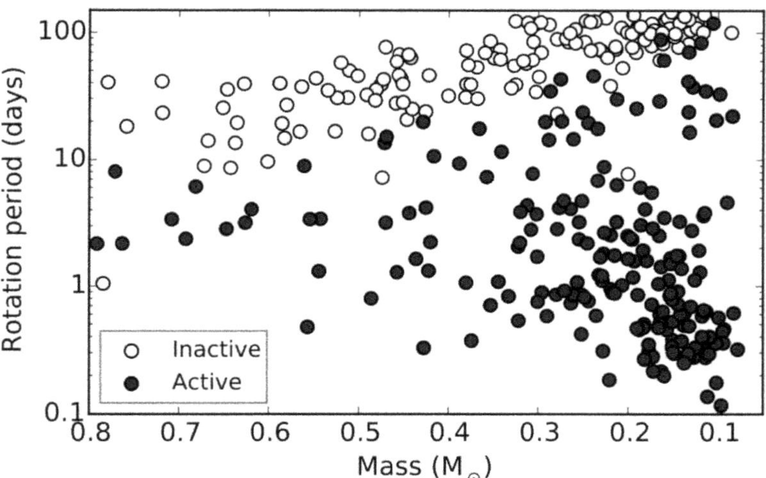

Figure 5.9. The photometric rotation periods of active (Hα emission equivalent width greater than 1 Å) and inactive M dwarfs as a function of stellar mass. Reproduced from Newton et al. (2017). © 2017. The American Astronomical Society. All rights reserved.

rotators and they are slower for lower masses. It also shows that a population of rapid rotators increases as the mass decreases and that these active stars come to dominate the sample at the lowest masses and shortest periods. This could be due to the transition from fast to slow rotator taking longer at lower masses but then happening quickly and more effectively.

Another take on this question notes that the fraction of stars exhibiting Hα emission is lower for more massive M dwarfs compared with mid to low mass M dwarfs (as can be seen in Figure 5.9). One can ask how this fraction of active stars behaves as a function of mass and time. It has been clear in a general sense that activity (defined this way) decreases over time, because the very active earlier M dwarfs in the field (flare stars) are preferentially fairly young. There is also evidence from cluster data that the activity fraction is lower in older clusters. West et al. (2008, 2011) greatly expanded our knowledge of this question by studying over 40,000 M dwarfs from the SDSS and 2MASS catalogs. This combined data set provides colors that can be calibrated to stellar mass, kinematics, and activity information (from spectra). The surveys are deep enough to capture not just nearby red dwarfs in the disk, but the older population (including spectral sub-types as late as M7) that is kinematically inflated to larger scale heights by long interaction times in the Galaxy. The study is aided by the fact that cooler stars also tend to have larger emission strengths, which helps compensate for the fact they are fainter at a given distance.

These studies confirmed trends from previous studies that the early M dwarfs have a fairly sharp drop-off in activity at moderate ages. Its rapidity is different than what

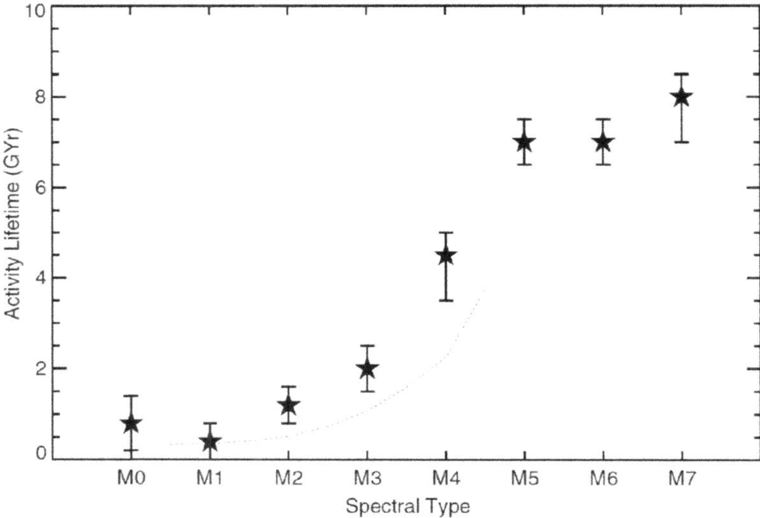

Figure 5.10. The length of time during which M dwarfs are active (as defined by their Hα emission) as a function of spectral type. The relation is the result of a study of more than 40,000 red dwarfs to heights well out of the local thin disk. Credit: Reproduced from West et al. (2008). © 2008. The American Astronomical Society. All rights reserved.

would be expected if activity decreased smoothly with time. The drop-off moves to larger ages as the stellar mass decreases. In order to understand their data they combined kinematic models of how stars "heat up" with time as they encounter other stars (primarily in the disk, where they also tend to form) with models of how the activity decreases with age. This combined analysis yields the result in Figure 5.10 that shows that the "activity lifetime" of early M stars is less than 2 Gyr, but increases from M3 to M7 up to about 8 Gyr. The same relations hold for the higher Balmer lines and Ca II when these more difficult observations can achieve the same sensitivity (West et al. 2011). This is important information if one is worried about the effect of red dwarf magnetic activity on exoplanets in their habitable zones, which are worryingly close to this type of star. The habitable distance moves rapidly closer as stellar mass decreases because stellar luminosity is dropping very quickly while the surface fluxes of UV and X-rays are not decreasing nearly as quickly (sometimes not at all). The behavior of very low mass stars contains a number of additional puzzles that are discussed in Section 7.1.

References

Baliunas, S. L., & Vaughan, A. H. 1985, ARA&A, 23, 379

Barnes, S. A. 2003, ApJ, 586, 464

Barnes, S. A. 2010, ApJ, 722, 222

Basri, G., Marcy, G. W., & Graham, J. R. 1996, ApJ, 458, 600

Boro Saikia, S., Marvin, C. J., Jeffers, S. V., et al. 2018, A&A, 616, A108

Bouvier, J., Matt, S. P., Mohanty, S., et al. 2014, Protostars and Planets VI, ed. H. Beuther, R. S. Klessen, C. P. Dullemond, & T. Henning (Tucson, AZ: Univ. Arizona Press), 433

Browning, M. K. 2008, ApJ, 676, 1262

Chaplin, W. J., Basu, S., Huber, D., et al. 2014, ApJS, 210, 1

Finley, A. J., & Matt, S. P. 2017, ApJ, 845, 46

Gizis, J. E., Reid, I. N., & Hawley, S. L. 2002, AJ, 123, 3356

Godoy-Rivera, D., Pinsonneault, M. H., & Rebull, L. M. 2021, arXiv:2101.01183

Gordon, T. A., Davenport, J. R. A., Angus, R., et al. 2021, ApJ, 913, 70

Gruner, D., & Barnes, S. A. 2020, A&A, 644, A16

Jeffers, S. V., Schöfer, P., Lamert, A., et al. 2018, A&A, 614, A76

Karoff, C., Metcalfe, T. S., Santos, Â. R. G., et al. 2018, ApJ, 852, 46

Kim, Y.-C., & Demarque, P. 1996, ApJ, 457, 340

Kiraga, M., & Stepien, K. 2007, AcA, 57, 149

Kraft, R. P. 1967, ApJ, 150, 551

Lehtinen, J. J., Spada, F., Käpylä, M. J., Olspert, N., & Käpylä, P. J. 2020, NatAs, 4, 658

Luyten, W. J. 1926, BHarO, 835, 2

Mamajek, E. E., & Hillenbrand, L. A. 2008, ApJ, 687, 1264

McQuillan, A., Mazeh, T., & Aigrain, S. 2014, ApJS, 211, 24

Metcalfe, T. S., & Egeland, R. 2019, ApJ, 871, 39

Mittag, M., Schmitt, J. H. M. M., & Schröder, K. P. 2018, A&A, 618, A48

Newton, E. R., Irwin, J., Charbonneau, D., et al. 2017, ApJ, 834, 85

Noyes, R. W., Hartmann, L. W., Baliunas, S. L., Duncan, D. K., & Vaughan, A. H. 1984, ApJ, 279, 763

Pizzolato, N., Maggio, A., Micela, G., Sciortino, S., & Ventura, P. 2003, A&A, 397, 147

Popinchalk, M., Faherty, J. K., Kiman, R., et al. 2021, ApJ, 916, 77

Radick, R. R., Lockwood, G. W., Henry, G. W., Hall, J. C., & Pevtsov, A. A. 2018, ApJ, 855, 75

Reiners, A., Schüssler, M., & Passegger, V. M. 2014, ApJ, 794, 144

Reinhold, T., Shapiro, A. I., Solanki, S. K., et al. 2020, Sci., 368, 518

Schöfer, P., Jeffers, S. V., Reiners, A., et al. 2019, A&A, 623, A44

Skumanich, A. 1972, ApJ, 171, 565

Sood, A., Kim, E.-j., & Hollerbach, R. 2016, ApJ, 832, 97

Spada, F., & Lanzafame, A. C. 2020, A&A, 636, A76

van Saders, J. L., Ceillier, T., Metcalfe, T. S., et al. 2016, Natur, 529, 181

van Saders, J. L., Pinsonneault, M. H., & Barbieri, M. 2019, ApJ, 872, 128

Vaughan, A. H., & Preston, G. W. 1980, PASP, 92, 385

West, A. A., Hawley, S. L., Bochanski, J. J., et al. 2008, AJ, 135, 785

West, A. A., Morgan, D. P., Bochanski, J. J., et al. 2011, AJ, 141, 97

Wilson, O. C. 1963, ApJ, 138, 832

Wright, J. T., Marcy, G. W., Butler, R. P., & Vogt, S. S. 2004, ApJS, 152, 261

Wright, N. J., Newton, E. R., Williams, P. K. G., Drake, J. J., & Yadav, R. K. 2018, MNRAS, 479, 2351

An Introduction to Stellar Magnetic Activity

Gibor Basri

Chapter 6

Stellar Magnetic Fields

What is clear about the magnetic field on a star like the Sun is that it cannot be primordial (the Ohmic dissipation times are extremely short compared to its age) and some means of regenerating field is obviously operating because of the solar activity cycle and its reversing polarities. Magnetic field can be regenerated when three ingredients are present: a conducting fluid, turbulent motions (like convection), and global rotation. All stars are made of conducting plasma and all stars rotate. The stars we have been discussing so far have outer convective envelopes or are fully convective, but even stars with radiative envelopes have convective cores and can sometimes exhibit surface magnetic fields (Section 7.4).

An extensive review of the physics and methods of measurement directly relevant to this book has recently been written by Kochukhov (2021; hereafter KoRev21). The discussion and figures in KoRev21 provide such an excellent exposition of the topics covered here that I often refer the reader to that source to gain a more detailed introduction. That review concentrates on recent developments and observations relevant to M dwarfs, however, while this book includes solar-type and hot stars and a fuller historical treatment.

6.1 Magnetic Dynamos

There are two fundamentally different approaches to the theory of magnetic dynamos that could crudely be characterized as concentrating more on the analytic side or more on the computational side. That does not mean that either is free of the other approach however. The more analytic approach is called "mean field theory" and involves many fundamental approximations. The second approach is to use MHD theory in as much detail as is computationally possible and run numerical simulations with as great a resolution and volume coverage as possible. These can be extremely computationally intensive and also involve many approximations as well as not being able to reach some parts of parameter space that stars operate in.

doi:10.1088/2514-3433/ac2956ch6

The idea behind mean field theory is to think of the production of magnetic fields by dynamos as taking place on different scales. The small scale is set by the velocity field responsible for producing currents that generate magnetic fields on the scale of the flows (often this scale is the convective scale or pressure scale height). Small-scale dynamos can operate on this scale through what is commonly designated as the "α" effect. Thus α^2 dynamos are small-scale dynamos that depend mostly on convective overturn and produce fields that appear on the scale of granules or supergranules. On the other hand there can be dynamos that operate on much larger scales and rely more heavily on Coriolis forces for their functioning, called $\alpha\Omega$ dynamos (where Ω stands for the bulk rotation). The truth is that this is a somewhat artificial distinction; a large-scale dynamo can have strong small-scale fields within it and small-scale dynamos can sometimes end up producing an amount of large-scale field.

Mean field theory averages the small-scale motions on intermediate scales, meaning scales much larger than the small scale and much smaller than the large scale. It condenses these averages to a few coefficients and only tries to calculate the mean large-scale fields that will be generated. There is a great deal of complexity and judgment involved in deciding how to proceed with the averaging and what approximations to make, so there are many different mean field calculations. A sample of a mean field calculation relevant to the solar $\alpha\Omega$ shell dynamo (taking advantage of the shear between the radiative and convective layers) can be found in Küker et al. (2001). This type of cyclic dynamo relies on the conversion of a poloidal field to a toroidal field by latitudinal differential rotation and then a conversion of toroidal field back to poloidal field by some sort of dissipative processes. Once stars are fully convective a shell dynamo cannot operate, but even so it is possible to construct mean field dynamos that produce strong fields. An example relevant to α^2 dynamos in fully convective objects was presented by Chabrier & Küker (2006) and later work is discussed in KoRev21.

Numerical simulations have been done for decades to try to understand both the Earth's and Sun's dynamos. It is not particularly difficult to generate a field in either case, even a field with about the right strength and filling factor. What is hard is to reproduce is the timescales over which the the field undergoes reversals, and in the Sun's case the way that the reversal unfolds (the butterfly diagram; see Section 6.1.1). For a long time it appeared that the best models for the solar dynamo required it to operate primarily at the tachocline, the interface between the radiative core and the convective envelope. A whole class of models called "interface dynamos" were developed, although they proved somewhat finicky. Numerical simulations began to call the primacy of the tachocline into question after the turn of this century, and then the stellar observations described in Section 6.2 made it clear that stars lacking a tachocline altogether still can produce strong and large-scale fields that depend on their rotation rate. The primacy of the tachocline has now fallen out of favor.

Numerical simulations must also make a number of critical approximations and assumptions. They can model critical ingredients like convection and rotation reasonably well, but cannot computationally accommodate the crucial magnetic

Reynolds number within orders of magnitude of its actual values in stars. That is a dimensionless variable that compares induction or advection with diffusivity on a given length scale. Because the diffusivity is very small on the relevant scales, R_m is very large. There are various treatments of subgrid scales, but the problem encompasses far too large a range in both spatial and timescales to fully treat. All simulations have to use strongly enhanced numerical dissipation methods for viscosity, thermal diffusivity, and magnetic diffusivity. The surface boundary conditions are too complex and difficult to treat so the simulation must stop somewhere below. One approach that has made relatively more realistic computations for both the Sun (Browning et al. 2006) and fully convective stars (Browning 2008; Yadav et al. 2015, 2016) are the anelastic spherical harmonic codes ASH and MagIC. These references provide discussion of some of the issues that numerical models struggle with, and an example of what they produce is shown in Figure 6.1. An alternative method that includes some relevant physics and is somewhat intermediate between the analytic models and the full resolved numerical treatment is the flux emergence approach of Isık et al. (2018) and related papers.

An excellent, fairly lengthy but readable critical review of solar dynamo theory has been provided by Charbonneau (2010). This is where the reader who wants more detail and equations should start (despite initial appearances the article is open access). He discusses the evolution of thinking and work on dynamos with critical assessments of a number of topics. In particular, the Babcock–Leighton (BL) model first presented in the 1960s fell out of favor in the 1970s in favor of more electrodynamical ideas, but came back into favor as those ran into theoretical trouble and observations seemed to confirm the basic idea behind BL. It was inspired by the observation that large bipolar sunspot pairs tend to emerge with leading spot at a slightly lower latitude than the trailing spot, this "tilt" is reduced as the spot pairs emerge closer to the equator in the later part of the cycle (this behavior is known as "Joy's law").

The tilt of the magnetic axis of emerging bipolar field implies a net dipole moment on the N–S axis, and as the spots decay and their field is dispersed by surface flows a fraction of this net dipole can end up contributing to the global dipole (increasing the poloidal field). Observations supporting this mechanism are presented in Section 6.1.1. Only 0.1% of the flux that emerges in bipolar active regions is needed to account for the average polar cap flux. This BL type of model is called advection-dominated if it requires surface meridional flow to transport the flux poleward, as has been observed (Hathaway & Rightmire 2010). Deeper return flows are needed to explain the equatorward drift of the general sunspot emergence locations. Meridional flows also appear crucial to explain the inner differential velocity patterns in the convection zone revealed by helioseismology, but they are slow and hard to measure well (Hathaway et al. 2003). An interesting idea (called a "solar tsunami") about how the next cycle is actually triggered has recently been proposed by Dikpati et al. (2019) although it is not yet the accepted explanation. Various dynamo models achieve various parts of the solar constraints on timing and cycle evolution, but none are perfect. Charbonneau (2010) discusses important then remaining issues.

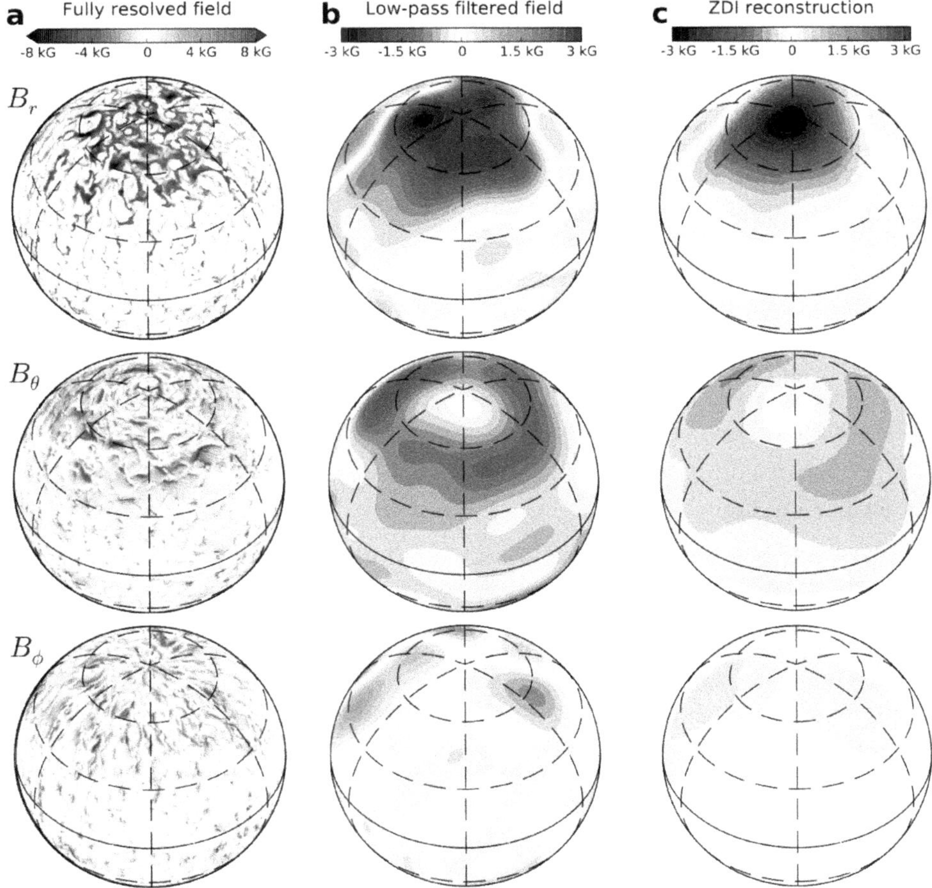

Figure 6.1. Results from an anelastic simulation of a rapidly rotating M dwarf by Yadav et al. (2015). Three components of the field are shown on the left with polarity (red and blue). The simulation actually has even higher resolution. The middle panel shows the same field at low resolution, and the right panel shows that what ZDI (Section 6.2.2) would detect appears similar. Credit: Reproduced from Yadav et al. (2015). © 2015. The American Astronomical Society. All rights reserved.

For fully convective stars the important conclusion seems to be that the α effect is able to generate magnetic fields until the magnetic energy is comparable to the kinetic energy. There is evidence that the α effect also works in the Sun at some level, providing the non-cyclic part of the solar field. On M dwarfs the surface field grows to several kilogauss covering much of the star (Section 6.2.1) and can be either axisymmetric or not (Section 6.2.2). Because of the near equivalence of magnetic and kinetic energies differential rotation is strongly suppressed. In principle this can be tested observationally but in practice it has not yet been done convincingly.

Radiation actually carries a significant part of the energy in the inner part of the star despite it being fully convective so the convective cells are slow and large and this allows rotation to play a greater role (Browning 2008). Nearer the surface convective cells become smaller and more vigorous. Whether the total magnetic flux would retain a clear dependence on rotation for non-solar-type dynamos was one of the main questions that both modeling and observations sought to answer in the 1990s. Recent models clearly exhibit a dependence on rotation and observations in the same time frame reached the same conclusion (Reiners 2012).

The structure of the field is more poloidal for faster rotation and more complex and non-axisymmetric for slower rotation (Yadav et al. 2016). Low resolution versions of these simulations can resemble images made from polarization techniques (Figure 6.1) as described in Section 6.2.2. That section also points out, however, that there appears to be a lot of additional magnetic field on real stars that is not detected in polarization observations because opposite polarities cancel out each other unless sufficiently separated by Doppler shifting due to stellar rotation. The co-existence of large- and small-scale magnetic fields is addressed by Yadav et al. (2015), among others.

6.1.1 Stellar Cycles

The solar cycle has been mentioned many times above in various contexts. It was discovered in the 18th century; the 1755 cycle is numbered "1" and Cycle 25 began in Dec. 2019. It was noticed through the number of sunspots that appear over rotational averages but more basically is a cycle in the emergence and polarity of the solar magnetic field. There is also a pattern to the latitude where sunspots typically emerge, as shown in Figure 6.2. Sunspot numbers rise and fall with an average 11 year period (varying between 9 and 14 years) but the polarity of the polar

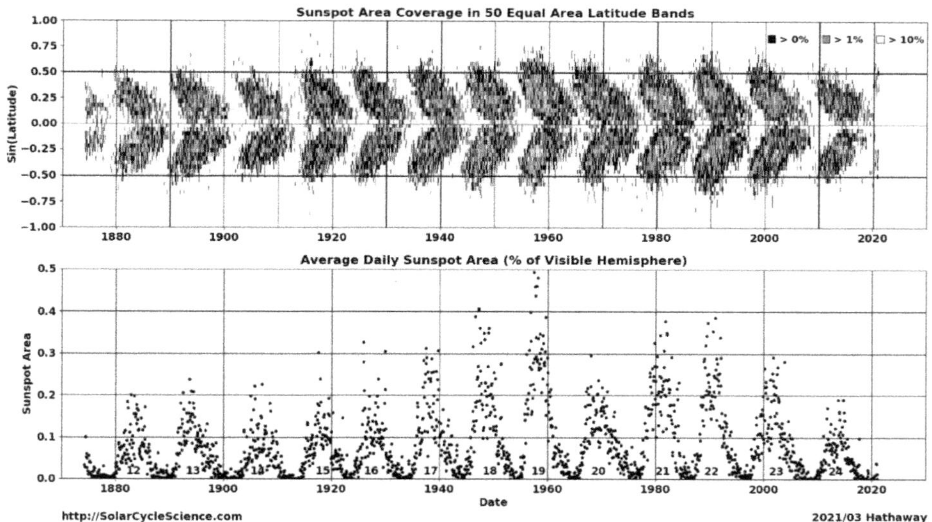

Figure 6.2. The so-called "butterfly diagram" showing the locations (upper panel) and amplitudes (lower panel) of sunspots over a number of solar cycles. Note that the spots emerge at a latitude around 30 degrees at the beginning of each cycle then emerge at progressively lower latitudes as the cycle progresses. Each cycle has the opposite polarity of the last (Figure 6.3). Credit: Reprinted with permisison from David Hathaway.

caps and orientation of bipolar regions on opposite sides of the equator reverse themselves by each minimum so the true cycle period is 22 years (Figure 6.3). In addition to sunspot number the coverage of active regions follows similar patterns and the brightness of the X-ray corona can increase by several times from cycle minimum to maximum. These phenomena reflect the underlying behavior of the magnetic dynamo that keeps the Sun's magnetic field active.

Both the periods and amplitudes of the solar cycles vary somewhat (Figure 6.2). There have also been cases of "Grand Minima" in which almost no sunspots are seen. The best known of these cases is the "Maunder Minimum" that lasted from 1645–1715, encompassing what should have been several cycles. That is based on visual observations of sunspots but it is possible to reconstruct the history of previous cycles through indirect proxies. Some of these are related to the high energy particles that the Sun directly sends our way or the changing partial protection it offers from galactic cosmic rays due to the solar wind. Isotope levels produced by cosmic rays and stored in tree rings provide high time resolution, and the theory of the production rates has been tested against modern data of all the relevant factors (solar activity, geomagmetic variations, climate variations). One such reconstruction is that of Usoskin et al. (2021) using the production of radioactive ^{14}C to infer levels of solar activity for the last thousand years. This data shows that the Spörer Minimum (1410–1540) was even longer than the Maunder Minimum. Other methods (with lower accuracy) have been extended back ten thousand years and more. These suggest that the Sun has spent twice as much time in the past millennium less active than recent levels. Apparently we are lucky it has been fairly active (in the top ten percent) for study by modern instrumentation. A good recent review of what is known observationally about the solar activity cycle has been given by Hathaway (2015).

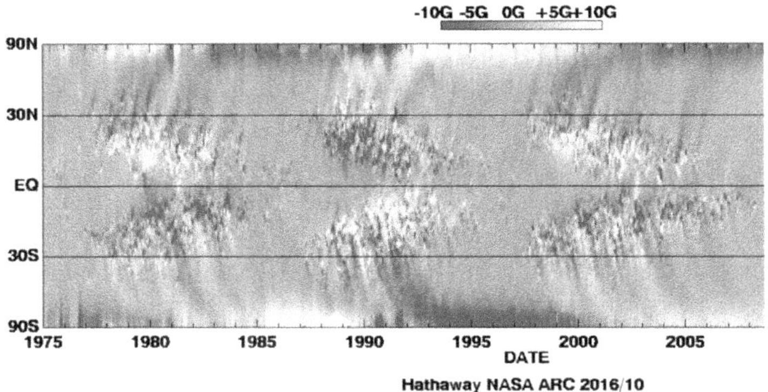

Figure 6.3. Another version of the butterfly diagram, this time with polarization information. It shows the radial magnetic field with time and latitude. Notice the drift of field with the opposite polarity of the dominant emerging field in each hemisphere that drifts poleward and eventually reverses the polarity of the polar cap. Observations of this type support some aspects of the BL dynamo model. Credit: Reprinted with permisison from David Hathaway.

We are interested in studying activity cycles on other stars for a number of reasons. We would like to know how cycle periods and amplitudes are related to stellar mass, age, and activity levels, how they are related to field configurations and symmetries, and under what circumstances the magnetic activity is cyclic or varies randomly. This information would allow a better connection between dynamo theories and surface phenomena and provide a way to test the processes we think are taking place inside the stars related to magnetic field production, transport to the surface, and dissipation. The primary study of cycles on other stars is the Mt. Wilson Survey (Baliunas et al. 1995). This used the S-index for Ca II described in Section 5.1 and covered 111 main sequence G and K stars (some were later found to be subgiants) from 1966 to 1994 although not all stars were observed for the whole interval. The data were searched for periods using Lomb–Scargle periodograms and their false alarm probabilities used to assign a quality to the cycle detection. A sample of the data is given in Figure 6.4.

A number of cycles were found, some very clear (especially in cooler stars), with periods comparable to or shorter than the Sun's. It would have been hard to find periods much longer than the Sun's, but some of the stars showed systematic long-term trends that could possibly be long cycles. Baliunas et al. (1995) divided the sample into active and inactive halves based on the scatter of S-indices at each color. The cycle statistics for these two groups are shown in Figure 6.5. They show that about half of the stars had detected cycles. Active stars have shorter periods on average and a larger fraction of chaotically variable cases. Less active stars showed fewer cases of chaotic variability, the only cases of no variability, and equal numbers of long-term trends. Another program performing long-term S-index surveys is the Lowell Observatory SSS program (Hall et al. 2007). A new large sample of S-index cycle determinations is about to be published as this is written by the California Planet Survey (CPS) group and no doubt similar studies will be done by the HARPS groups. Exoplanet searches in general are paying more attention to stellar activity because it can affect radial velocity signals at the precision now being achieved.

Several attempts to infer dynamo properties from the Mt. Wilson data have been made. One example is by Baliunas et al. (2006) and a more recent paper is Brandenburg et al. (2017). The latter paper brings in some data from other sources and speculates about dynamos with concurrent short and long periods. I cautioned in Section 2.2 that short periods from differential space photometry might not reflect true magnetic cycles but simple rearrangements of starspot distributions. This caution applies to the large number of stellar activity cycles claimed by Reinhold et al. (2017), but that paper provides a good list of references of work on stellar cycles to that date. It is more likely that cycles can be detected given absolute photometry because faculae then contribute a large signal and seem to provide better information due to their larger total area. Examples of such cycles were shown by Montet et al. (2017). If the TESS spacecraft lasts long enough it might provide many other examples because it records full-frame images that hold the promise of absolute calibration. There are now a number of ground-based long-term photometric surveys that may also contribute in the future.

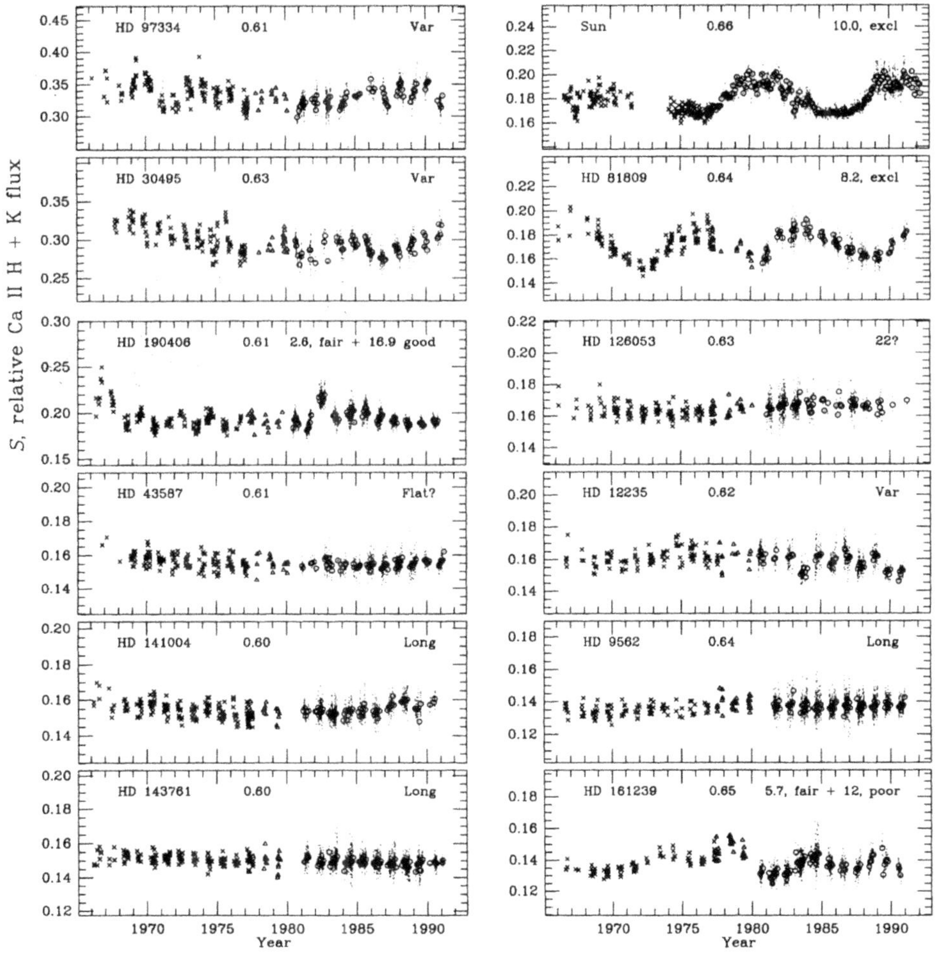

Figure 6.4. A sample of S-index activity cycles on solar-type stars from the Mt. Wilson Survey (including the Sun). Each panel contains the star name on the left, the B–V color in the middle, and a classification on the right. If a cycle was detected the period in years is given along with a quality flag (sometimes two periods are suggested). If not the data was classified as variable but not cyclic ("var"), not variable ("flat"), or showing a long-term trend ("long"). Credit: Reproduced from Baliunas et al. (1995). © 1995. The American Astronomical Society. All rights reserved.

Attempts have also been made to observe butterfly diagrams on other stars. That requires knowing the latitudes of starspots, which we found in Section 2.3 is not very easy. As mentioned there, the best accuracy and resolution can be obtained using planetary transits over starspots. This has been achieved for only a few stars so far, an example is provided by Netto & Valio (2020). Even in this case one may not be measuring the true spot distributions as demonstrated by Namekata et al. (2020).

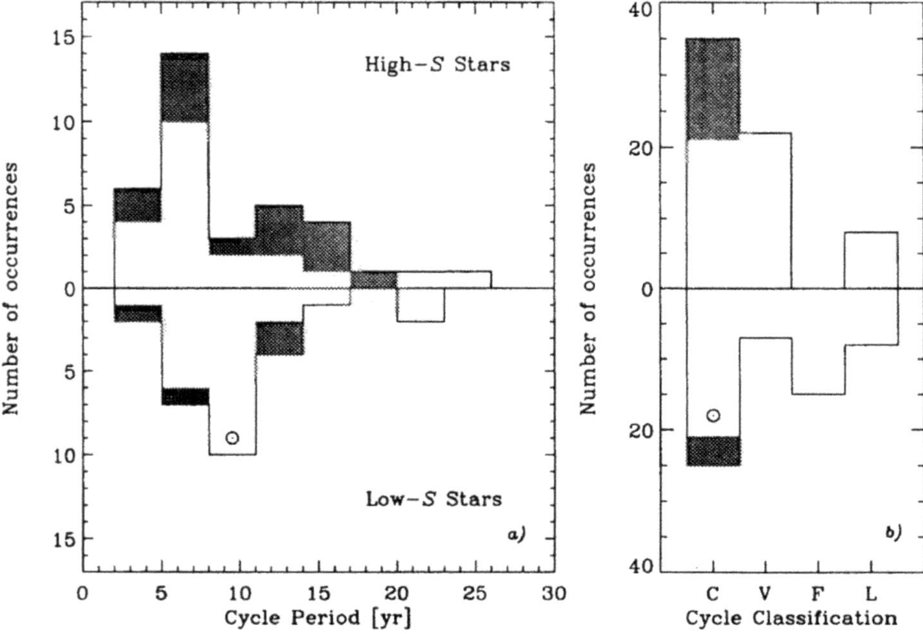

Figure 6.5. The distribution of cycle periods from the Mt. Wilson Survey in three year bins. The upper left histogram is for more active stars and the lower left is for less active stars. The shaded areas represent secondary periods also found. The right-hand histograms give the distribution of light curves, whether cyclic, variable, flat, or showing a long-term trend. Credit: Reproduced from Baliunas et al. (1995). © 1995. The American Astronomical Society. All rights reserved.

6.2 Direct Measurement of Magnetic Fields

To really understand stellar magnetic activity and learn about dynamos it is helpful to be able to measure the magnetic field itself, even if only for relatively few stars. This is more fundamental and different from making indirect inferences using proxies that rely on effects of the magnetic field on the stellar atmosphere such as non-radiative heating. Up to now we have discussed only these proxies. It is also important to keep in mind the difference between the magnetic field strength which varies over the surface, and the (total) magnetic flux which is usually taken as the product of the average field strength and the surface filling factor. The field strength varies over the stellar surface and its highest values are easiest to detect if they have sufficient filling factor and the regions containing them are not too dark. The physical basis for direct field measurements is the Zeeman effect, which arises from the fact that atoms have magnetic moments. In the presence of a magnetic field their interaction with it splits atomic levels into magnetic sublevels by coupling with the spin and orbital angular momenta. The amount of splitting depends on the magnetic field, but also on the quantum mechanical properties of the particular spectral transition under consideration. Lorentz and Zeeman won the 1902 Nobel Prize for discovering this effect.

The simplest case of the Zeeman effect (see Figure 1 in KoRev21) is when the atomic level splits into three components, with magnetic quantum numbers of 0, ± 1. The $m = 0$ component (called the π component) remains unshifted while the plus and minus components (called σ components) shift in opposite directions in wavelength. All three components also become polarized in ways that depend on the orientation of the magnetic field to the observer; the behavior of the π component is complementary to that of the σ components. A full characterization of the polarization state of the radiation requires knowledge of all four Stokes polarization parameters (cf Section 6.2.2). Depending on the structure of the atomic transition and the strength of the field, things can become significantly more complicated, and molecular transitions add a further layer of complications.

The measurement of magnetic fields on stars started with the observation of Zeeman splitting of lines formed in sunspots by George Ellery Hale in 1908. Figure 6.6 shows the sort of observation he made, and also demonstrates the principle behind measuring magnetic broadening of spectral lines on other stars. Since this method only employs spectral intensity it is making use of the Stokes I parameter. Modern observations of magnetic fields on the Sun are done with magnetographs (Figure 6.7), instruments that difference the image in positive and negative circular polarization to provide an easier means of measuring values for the longitudinal (toward the observer) magnetic field. These utilize the Stokes V parameter. An instrument that measures all four Stokes parameters is called a vector magnetograph, and it enables the full characterization of the direction of the magnetic field.

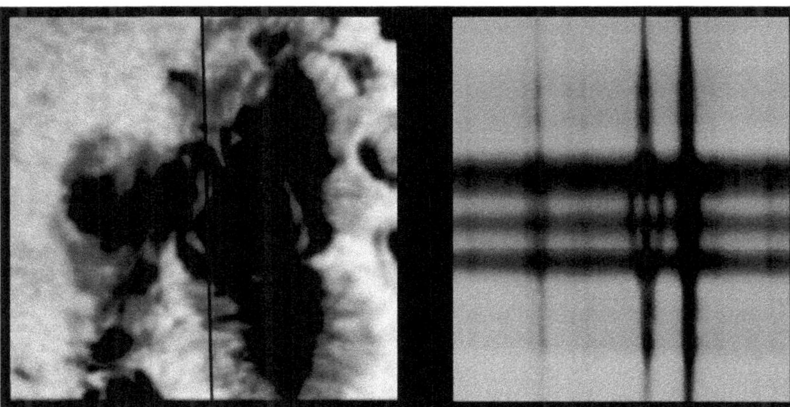

Figure 6.6. Zeeman splitting of a magnetically sensitive iron spectral line at 525 nm. The left (yellow) image shows a region including a sunspot; the thin vertical line near the middle is the spectrograph slit. The right image shows the spatially-resolved (vertical direction) spectrum (wavelength dispersed in the horizontal direction). A magnetically sensitive line splits into 3 Zeeman components in the part of the slit that samples the sunspot (halfway up the green image). The strength of the field in the center of the umbra is measured at a little over 4 kilogauss. Credit: Reproduced with permission from NSO/AURA/NSF.

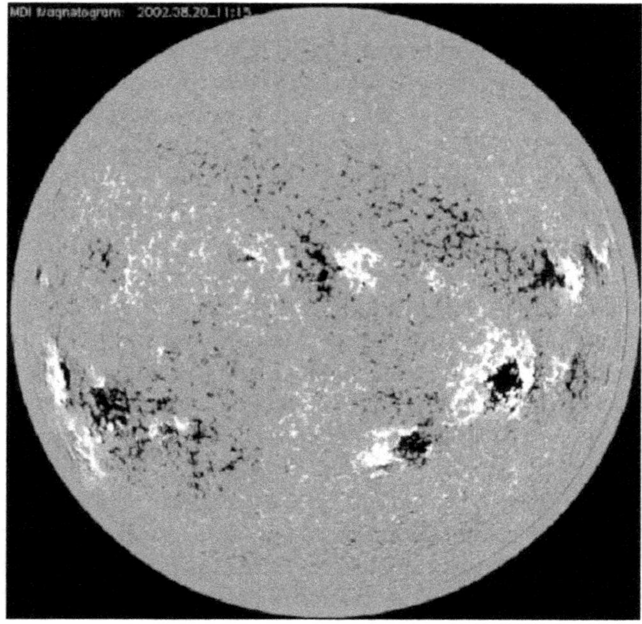

Figure 6.7. An image of the Sun taken with a magnetogram (the MDI instrument on SOHO). Opposite polarities are shown in black and white. The image was taken on 2002 Aug. 20 when the Sun was near the maximum of its activity cycle. Credit: Reprinted from Domingo et al. (2005), Copyright (2005), with permission from Elsevier.

6.2.1 Zeeman Broadening in Stars

KoRev21 goes through the derivation of the primary equation of interest in understanding observations of the Zeeman effect in stars, which expresses the change in wavelength of a spectral feature caused by a magnetic sublevel of a transition as

$$\Delta\lambda_\mathrm{B} = 4.7 \times 10^{-12} g_\mathrm{eff} B \lambda_0{}^2 \tag{6.1}$$

where λ is in nm and B is in Gauss. The factor g_eff is the effective Landé g-factor (see KoRev21 for its definition) that gives the coupling strength between the transition and the field. For example a neutral iron line at 846.8 nm that has been used for this purpose has magnetic sublevels with $g_\mathrm{eff} = 2.5$. A kilogauss field will produce a shift of the sublevels of 8×10^{-3} nm, while the Doppler shift of that line induced by turbulent motions of 2 km s^{-1} (typical for solar-type stars) is 5×10^{-3} nm. This renders the magnetic broadening of the line unmeasurable unless the average field on the star is at least several hundred Gauss. It also means that to see it one needs high spectral resolution (greater than 40,000) and high signal-to-noise (greater than 100). Because the magnetic broadening grows as the square of the wavelength while Doppler broadening grows linearly it is advantageous to go to longer wavelengths.

It is important to remember that the spectral changes in a line originating in magnetic field regions are diluted (often severely) by the signal from non-magnetic regions, since except for the Sun we are observing the disk-integrated spectrum. There will generally also be a distribution of field strengths mixed together. Nonetheless, Zeeman broadening is often simply interpreted as due to one field strength with a certain filling factor on the star. Because magnetic flux tubes expand above the photosphere one has to be a little careful interpreting filling factors; they should generally be expected to be smaller for diagnostics that are formed at greater depths, but the product of field strength and filling factor (integrated magnetic flux) is hopefully more constant.

Serious efforts to measure magnetic fields on solar-type stars using Zeeman broadening began in the 1980s. The first survey with tens of stars was done by Marcy (1984) who also summarized previous efforts to measure fields and the method-ologies in use (including polarization). The idea is to deal with the fact that stellar convective and turbulent motions produce spectral line broadening of the same order as strong magnetic fields by observing two lines from the same element with similar strengths and excitation temperatures so that their formation is in the same part of the stellar atmosphere and subject to the same motions. One of the lines should be significantly more sensitive to magnetic fields (have a higher Landé g-factor) than the other, so that additional broadening in the sensitive line can be ascribed to the field. The lines used in this early work are two neutral iron transitions at 617.3 and 624.1 nm whose g-factors are 2.5 and 1.0, respectively. He found excess broadening of the magnetically sensitive line on two-thirds of his small sample of stars with greater broadening on the more active stars. Very simplified radiative transfer was used to translate the broadening to a field strength and filling factor.

This technique was refined and expanded in the following years. The treatment of radiative transfer was made increasingly sophisticated and the recognition that magnetic parts of the atmosphere are different from non-magnetic parts began to be taken into account. For example, although starspots are likely to have the strongest fields they are also relatively dark and contribute much less to the integrated profile. Magnetic flux tubes might be brighter (and faculae are) but their atmospheric structure is different than the quiet photosphere. The advantage of moving to longer wavelengths also began to become more practical from an instrumental point of view. A summary of the this progress can be found in Basri et al. (1990) who incorporated all of these concepts and utilized the iron line mentioned above at 846.8 nm along with its less sensitive twin at 774.8 nm. They concluded that this Zeeman broadening methodology is subject to a number of systematic errors of unknown amplitude, so the inherent uncertainties in the derived values of magnetic flux are of order a factor of two or more.

A more rigorous test was conducted by Valenti et al. (1995) using the Kitt Peak Fourier Transform Spectrometer (FTS) to observe an even more magnetically sensitive (g=3) iron line at a substantially longer wavelength (1565 nm). For the first time the σ components were actually resolved from the π component in a solar-type star, allowing a more definitive separate measurement of the field strength and filling factor in the active solar-type star ε Eri. This star had been featured in a number of previous studies that all found rather large magnetic fluxes on it, consistent with its strong Ca II emission. The new measurement confirmed that the dominant field strength is around 1.5 kG but found a filling factor of around 10%, yielding a magnetic flux at least a factor of two smaller than most previous determinations. Unfortunately the FTS is a much slower instrument than an echelle spectrometer and thus is confined to very bright stars (two other relatively active stars observed showed no magnetic signal).

Valenti et al. (1995) concluded that Zeeman broadening determinations in the optical are difficult and not very trustworthy; that was reinforced by the later work of Rueedi et al. (1997). It remains true today that most solar-type stars have magnetic fluxes too weak to be measured well by any of our current techniques. It is therefore still quite helpful to use a proxy that correlates closely with magnetic fields to assess the level of field present on such stars. The Ca II H&K lines are known to be very tightly correlated with magnetic fields on the Sun (Schrijver et al. 1989), so they or similar diagnostics serve the purpose.

The first resolution of σ components had actually been achieved earlier with the FTS by Saar & Linsky (1985) using Ti I lines at about 2.2 microns on the flare star AD Leo. Although M stars are faint in the optical they are much brighter in the infrared. The lines that are useful for solar-type stars (like upper levels of neutral iron) are not useful for cool stars but low ionization species like titanium that are ionized in solar-type stars work well for cool stars. Saar & Linsky (1985) found the two flare stars they observed have easily separated σ components that imply very strong (2–4 kG) fields covering most of the stars (Figure 6.8). This is despite their relatively rapid rotation because of the advantage of going into the infrared. They note that a simple contemporaneous dynamo theory (Durney & Robinson 1982) predicted strong fields for such convective and rapidly rotating stars along with their high surface gravity that implies large equipartition pressures. Subsequent work has confirmed that the M dwarfs have the strongest magnetic fluxes and those rapidly rotating are nearly covered by such fields. That makes them the best-studied and well-determined examples of directly measured stellar magnetic fields among stars with convective envelopes.

During this time period another method for measuring strong fields that has much less stringent observational requirements was developed by Basri et al. (1992). It allowed the first successful attempt to directly measure whether magnetic fields on T Tauri stars are strong enough to support the magnetospheric accretion and jet models that seemed to explain emission line observations of those stars (Section 7.2). This method, sometimes called "Zeeman intensification," relies on the behavior of the equivalent widths (integrated areas or line strengths) of spectral lines in the presence of the Zeeman effect.

The classical behavior of the line equivalent width relies on the concept of "saturation" of a line within its Doppler core. In the simplest case, if one imagines slowly increasing the amount of an absorber in a classic plane-parallel stellar atmosphere, the line will grow deeper until line center optical depth unity reaches the part of the atmosphere where the temperature gradient flattens out. Parts of the line increasingly off exact line center will join this deepest intensity as the number of

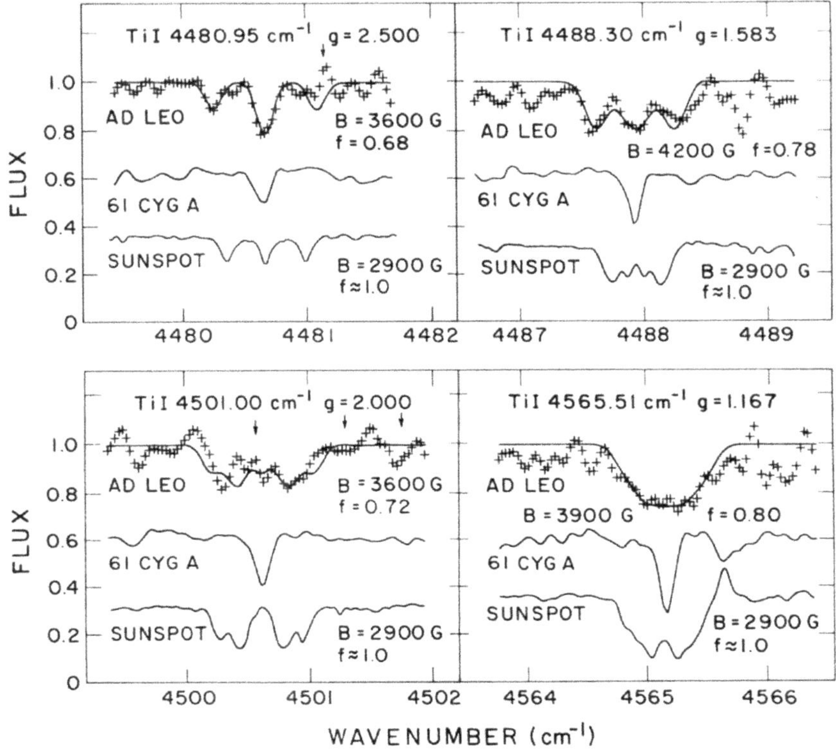

Figure 6.8. The first direct detection of stellar Zeeman splitting using Ti I lines near 2.2 microns from the dMe star AD Leo. Conversion of the observation to field strength and filling factor was done with radiative transfer modeling. Also shown are the spectra from a cool star that does not show the magnetic signal (61 Cyg A), and from a sunspot (which has a similar temperature). Credit: Reproduced from Saar & Linsky (1985). © 1985. The American Astronomical Society. All rights reserved.

absorbers continues to increase. However, the shape of the observed line will be dominated by Doppler shifting of individual line centers, forming the "Doppler core" of the line. Adding absorbers beyond this point does little to increase the integrated observed line strength because the individual absorbers just move around within the Doppler core and cover each other. Moving the integrated optical depth further out along a flat temperature gradient does not change the intensity at optical depth unity. The line wings are not yet strong enough to cause the continuum to be depressed outside the Doppler core. The observed line is "saturated"; only subtle changes near its bottom are occurring. If one keeps adding absorbers, at some point there will be sufficient added absorption that the line damping wings start to have significant opacity and depress the continuum outside the Doppler core. The equivalent width will begin to grow again at a different (faster) rate than the Doppler core did as the damping wings become deeper. This behavior is called the classical curve-of-growth (and also applies to absorption lines in the ISM).

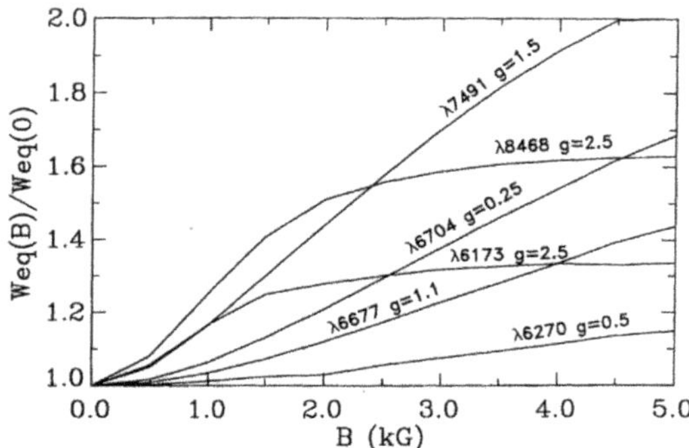

Figure 6.9. The expected growth of the equivalent widths of some iron lines in the red part of the spectrum under the influence of magnetic fields. Their effective Landé g-factors are given but their growth curves are not proportional to these. For example, due to its more complicated magnetic sublevels the 749.1 nm line is ultimately more useful than the 617.3 and 846.8 nm transitions despite their higher g-factors. Credit: Reproduced from Basri et al. (1992). © 1992. The American Astronomical Society. All rights reserved.

The behavior of the line strength is somewhat different if instead of adding absorbers the spectral transition contains magnetic sublevels and the strength of the magnetic field is increased. For a saturated line, so long as the Zeeman splitting is less than the Doppler broadening the equivalent width of the line will not change much because the line is already saturated. Once the Zeeman splitting exceeds this amount, however, the Zeeman components will move outside the Doppler core and begin to act like damping wings, although not with the same growth curve. Finally, once the components have cleared the Doppler core altogether they will simply continue to separate without adding more to the equivalent width because it is an integration over all relevant wavelengths.

Because transitions can have more complicated magnetic sublevels than a simple Zeeman triplet, and the strengths and g-factors for the different components can vary from each other, it requires a detailed analysis to determine how the equivalent width of a given line will grow with the magnetic field. An example of this is shown in Figure 6.9. The sensitivity of the lines to Zeeman intensification is not always intuitive. It may be more important to have more numerous mildly sensitive components than a few more sensitive components, and it is good to observe a set of lines with varying overall sensitivities to show that the line strengths all behave as they should if magnetic fields explain their strengths. One also has to have good values for the oscillator strengths of the lines; these are not always available or trustworthy and often require empirical calibration partly to the multi-component nature of stellar atmospheres. The advantage of this technique is that one does not need to make precise measurements of the line profiles; only the line strengths need be measured so signal-to-noise requirements are relaxed somewhat.

The methodology was not widely picked up after its initial use but now is enjoying a resurgence with much better instrumentation and the ability to go to longer wavelengths. Kochukhov & Lavail (2017) chose ten Ti I lines around 970 nm that show great promise for Zeeman intensification and they have been used along with other techniques in several papers since then. Muirhead et al. (2020) employ a variant of the method. Of course it is always true that operating at longer wavelengths is advantageous. The development of high resolution infrared spectrometers in the 1990s allowed more reliable measurements to be obtained, at least for the most active types of stars. In the mid 1990s it became possible to collect spectra of sufficient quality to study the magnetic fields on T Tauri stars in the infrared. Johns-Krull et al. (2000) review these and observations of active M dwarfs that use atomic line profiles. They all show high magnetic fluxes, with fields of several kilogauss covering much of the star.

Unsurprisingly, distributions of field strengths fit the observations better than a single field since that is likely the physical situation. Fits to line profiles that allow several different field strengths indicate that some fields on active M dwarfs and T Tauri stars lie well above the expected equipartition values with strengths above 5 kG (even up to 8 kG in extreme cases), although with rather small filling factors. Weaker fields are present as well and of course there are lower limits to what can be detected. Average fields and filling factors are what are usually reported but it is good to keep in mind that these are a crude representation of the true field present. More recent work uses more sophisticated treatment of the magnetic radiative transfer, including dropping certain assumptions (like a weak field approximation) that was used in earlier work. A summary to the present is in KoRev21.

As one moves to cooler stars and the M spectral class the atomic lines are increasingly masked by molecular lines and their best transitions not excited. Valenti et al. (2001) noted that lines in the Wing–Ford band of FeH around 990 nm show magnetic sensitivity in sunspots and might be useful for stellar measurements. This suggestion was adopted by Reiners & Basri (2007) who developed an empirical calibration using a few stars where both FeH and previously calibrated atomic lines could be tied together. This was then used in a series of papers to measure fields in about 70 M dwarfs, enough to establish correlations between the field and rotation periods and Rossby numbers. This work is summarized by Reiners (2012).

As mentioned in Section 6.1; one motivation was to test for a change in dynamo action or magnetic field production or sensitivity to rotation that could arise because the interface between a radiative and convective zone that was thought to be important for the solar dynamo could not be present once a star is fully convective (cooler than M3). We found that fully convective stars can have strong fields that still depend on rotation in similar ways to warmer stars (see Figure 11 in KoRev21 for an updated result). Saturation of the magnetic flux itself occurs at low Rossby numbers, which implies that saturation is not due to volume filling effects in the upper atmosphere. This conclusion is reinforced by Muirhead et al. (2020) who point out that these methods are measuring the magnetic flux in the photosphere. Later work involving infrared spectral synthesis found that the original method was fairly good for the earlier M stars it was calibrated on but can underestimate the field by up to a factor of two on the coolest stars. It became clear that the the magnetic fluxes on active M dwarfs are generally very large as first found by Saar & Linsky (1985).

A number of other papers by several groups have worked on the actual physics and radiative transfer for FeH and other molecules to better tie the observations to theory. Molecular Zeeman measurements will play an increasing role in the future for cool stars. Uncertainties still remain for each method (half a kilogauss or more) and different authors utilize different assumptions about atmospheric components and field distributions as well as varying levels of sophistication in the radiative transfer modeling. Nonetheless the broad outlines of how magnetic fields behave on cooler stars are relatively in hand. A good overview of all this work is in KoRev21.

The fact that low-mass stars can generate such large amounts of magnetic flux may even have a direct influence on their stellar structure. As measurements of stellar size have gained in accuracy and precision, a mismatch between predictions of stellar structure models and observations developed and has continued to be reinforced by later and more extensive work. The radii of some M dwarfs appear to be larger than any standard models predict. These inflated radii tend to occur in stars that exhibit strong magnetic activity, so theories have been proposed that directly connect strong interior fields to changes in convective behavior that will inflate the stellar radius (and concomitantly reduce the effective temperature). A recent relevant observation and good review of the preceding literature can be found in Lubin et al. (2017).

6.2.2 Magnetic Polarization in Stars

In addition to causing the separation of magnetic sublevels in atomic and molecular transitions, the presence of a magnetic field also introduces polarization into the radiative transfer. Polarization is the inducement of preferred directions in the EM vectors for photons. The usual way of describing polarization in this context is the use of the four Stokes components of the radiation vector: [I,Q,U,V]. I is the total intensity that we have been talking about in the last section. It is the quantity that is measured by a regular spectrometer as a function of wavelength. In order to utilize the polarization signal from a spectral line, the line must be resolved in wavelength as well as measured for polarization. The polarization signals are typically a small fraction of the total intensity.

To define the next two Stokes components one establishes a reference direction, for example North–South on the sky or some instrumental coordinate depending on how the observation is being done. The Q component is the difference between the amount of radiation with the electric vector oscillating along that reference direction and perpendicular to that direction. The U component is a similar difference but between electric vectors oscillating at a 45 degree angle to the directions used for Q. To measure these quantities, one must introduce an instrumental element that preferentially passes radiation oriented in those directions (polarizing filters). One must always be very careful to understand whether other optics in the full instrument (including the telescope) are introducing their own polarization signals. Finally, the V component is the difference between the signals passed through right-hand and left-hand circular (clockwise and counter-clockwise) polarization filters.

Let us first consider the case of a simple Zeeman triplet. When the field is pointing at the observer, the Stokes V signal has a characteristic hump-dip shape if the σ components are split out far enough (the red lines in the top row of Figure 6.10). It is

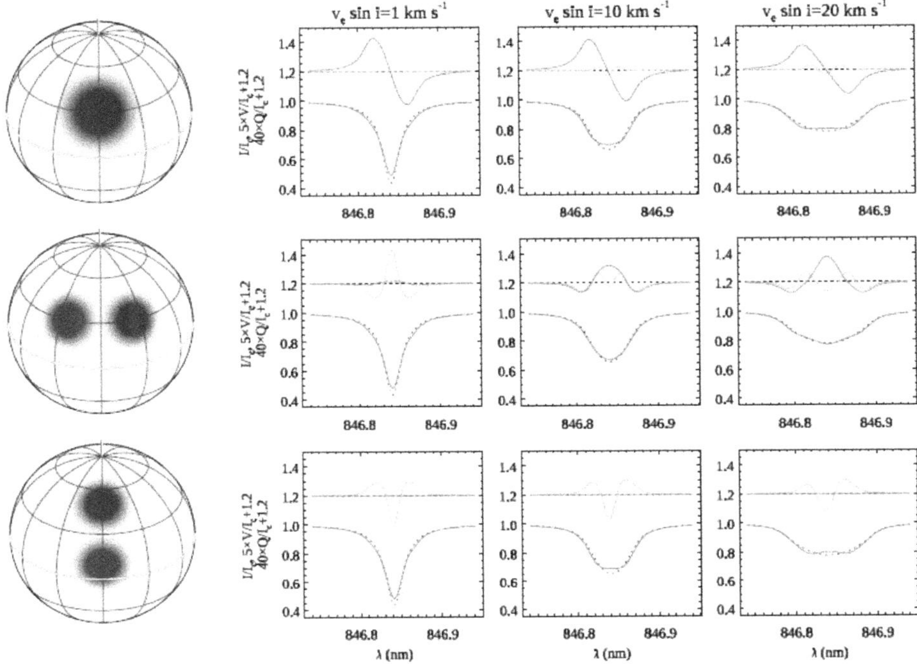

Figure 6.10. A set of illustrative examples of magnetic line profiles from Fe I 846.8 nm in various Stokes components from KoRev21. The field configurations are shown to the left of the plots—spots with 3 kG fields and opposite polarities in red and blue. The blue profiles on the right are for intensity (Stokes I), the red profiles are for Stokes V and the green profiles are for Stokes Q. The intensity profile in the absence of a magnetic field is given by the dotted lines. Each row shows three examples of Doppler broadening for 1, 10, and 20 km s^{-1}. Credit: Reprinted by permission from Springer Nature: Kochukhov (2021), © 2021, The Author.

this signal that magnetographs use to produce images like Figure 6.7. The circular polarization signal is reduced if the field is pointing at an angle to the observer and disappears if the field is perpendicular to the line of sight, while the linear polarization π component grows in the center of the line. The hump and dip exchange places when the magnetic polarity is reversed. This is the source of the black/white signals in a spatially-resolved magnetogram. It is a serious problem for magnetic observations of stars, however, which integrate the signal over the whole disk. The positive and negative polarities cancel in the integrated Stokes V signal so one can only measuring the amount of the net remaining polarized field or "uncancelled" field. Integrating this field over the surface yields the uncancelled flux. This can be quite a small proportion of the total flux if most of the field is largely in bipolar regions that are small compared to the stellar radius (as in Figure 6.7). The problem can be somewhat alleviated if different regions on the star are separated by rotational Doppler shifts.

Stokes Q has different behavior than V (green lines in Figure 6.10). The σ components grow in the same direction as each other and in the opposite direction of the π component, and do not cancel with opposite polarities. The signal is largest with a transverse field. Stokes U is similar but sensitive to different orientations. All the Stokes profiles except for V are symmetric about line center; V is anti-symmetric. Complications ensue when there are different polarities in different directions or a difference in the amounts of polarities, when the field is unevenly distributed around the visible hemisphere, when there is a distribution of field strengths, and due to Doppler shifting from stellar rotation. In order to understand and make physical interpretations from these observations one must perform a radiative transfer calculation that includes all four Stokes components through the atmospheric structures. This is possible with increasing sophistication, although we don't really know what atmospheric structures to use for active stars.

KoRev21 section 2.2 provides a summary of the state of affairs by 2020. Figure 6.10 shows three illustrative cases—a single polarity, two polarities on opposite sides of the stellar meridian, and two polarities along the stellar meridian. In the upper simplest case, the V signal is maximal and simple because the field is primarily radial and unipolar, but linear polarization suffers cancellation because the transverse field is pointing at all angles. Complications arise when stellar rotation shifts the profiles on the approaching and receding sides of the visible hemisphere. The middle row shows that V isn't fully canceled because of this at 5 km s^{-1} and the π component becomes strong at 20 km s^{-1}, while cancellation is complete in the bottom row because there is essentially no Doppler shift between the spots. See KoRev21 for a fuller explanation of other effects. Finally, the blue and dotted lines in Figure 6.10 show that the magnetic field always has an effect on I in this $g = 2.5$ 846.8 nm iron line (discussed in the previous section) but the effect requires very high precision to detect at these levels of magnetic filling factor and is harder to detect in more rapid rotators (however they tend to also have higher filling factors in reality).

An advantage of echelle spectrometers is that they simultaneously gather a large number of line profiles under the same observing conditions at the same time. Techniques that utilize this information to construct a composite line profile from many similar lines have been developed for a number of applications, including precision radial velocity searches for exoplanets. This has the advantage of averaging over a lot of individually less precise profiles to construct one that has much greater signal-to-noise. Radial velocity shifts of 0.001 pixels can be inferred from such composites that would be quite impossible for individual lines. A similar approach is advantageous in extracting information from spectropolarimetry. The technique in wide use for this purpose is called Least Squares Deconvolution (LSD). It was developed by Donati et al. (1997) and has been greatly refined since.

A carefully chosen set of atomic lines are cross-correlated with a set of delta functions having the same central wavelengths and amplitudes. This yields the average profile, which is treated as though it were the observed Stokes component. Until recently this was only done for Stokes V but efforts have begun to include Stokes I so that both the full and uncancelled magnetic flux are measured. The polarization information can be used to produce a magnetic field map of the large-scale asymmetries in the field. LSD starts to break down under various conditions (one example is fields greater than 2 kG) and for various lines so it becomes necessary to apply the LSD technique to calculated profiles in the same way as to observations to more accurately infer what the final signal means. For more details see KoRev21.

The technique of Doppler Imaging of starspots was presented earlier (Section 2.3); it also relies on averaging many lines together. That technique can be combined with LSD to produce a "Zeeman Doppler Image" (ZDI). This requires that a time series of spectropolarimetric measurements be made so that the whole of the star can be sampled and field concentrations can be viewed at different Doppler shifts. A map can then be made of the field on the stellar surface to the extent that opposite polarities do not cancel each other because they are at sufficiently large relative Doppler shifts. The usual method is to characterize the magnetic field with spherical harmonics; the low order harmonics represent the largest-scale structures.

ZDI provides a measure of the poloidal and toroidal components of the uncancelled field as well as the degree of axisymmetry. These are useful for evaluating some aspects of dynamo models or assessing what kind of angular momentum loss might happen via stellar winds. The latter is true because smaller bipolar structures are likely to be the source of lower-lying high-density closed coronal loops while larger more dipolar structures are likely to act more like coronal holes and produce much of the wind (Section 4.4). It also matters to what extent the wind comes out along the rotation axis (which removes little angular momentum) as opposed to more equatorial directions (Vidotto et al. 2014; Finley & Matt 2017).

An example of a ZDI map from Morin et al. (2010) of the flare star GJ 1156 in 3 different years is shown in Figure 6.11. ZDI can also be applied to slower rotators (although it no longer really utilizes Doppler information) using essentially only the temporal information. This produces very low spatial resolution on the star and suffers more from polarity cancellation. The premier instrument for gathering the requisite observations has been ESPaDOnS (Echelle SpectroPolarimetric Device for the Observation of Stars) on the 3.6 m CFHT (Canada France Hawaii Telescope) in Hawaii. Newer instruments are coming on line at larger telescopes that will make this sort of study feasible on more stars (if the substantial investment in observing time that is needed is granted).

The conclusions a decade ago from ZDI studies were summarized by Donati & Landstreet (2009) and are represented in Figure 6.12. Most obvious is the fact that low-mass stars have larger uncancelled fluxes that are in simpler configurations and tend to be dominated by toroidal fields. These are also stars with small Rossby numbers. This is consistent with theoretical expectations that rapid rotation will cause the field produced deep in the star to rise up on cylinders aligned with the rotation axis (Section 6.1) yielding predominantly poleward spots. One caution is that the slowly-rotating M dwarfs are absent from this sample. Higher mass stars have the opposite characteristics, and the solar-type stars with the slowest rotation (largest Rossby numbers) are barely detected. Perhaps most interesting effect is the possible transition between axisymmetric and non-axisymmetric uncancelled field configurations at Rossby number of about 0.1 for stars near 0.5 solar masses.

Note that most solar-type stars show up as having simple uncancelled field structures because almost all their flux has been canceled due to being from

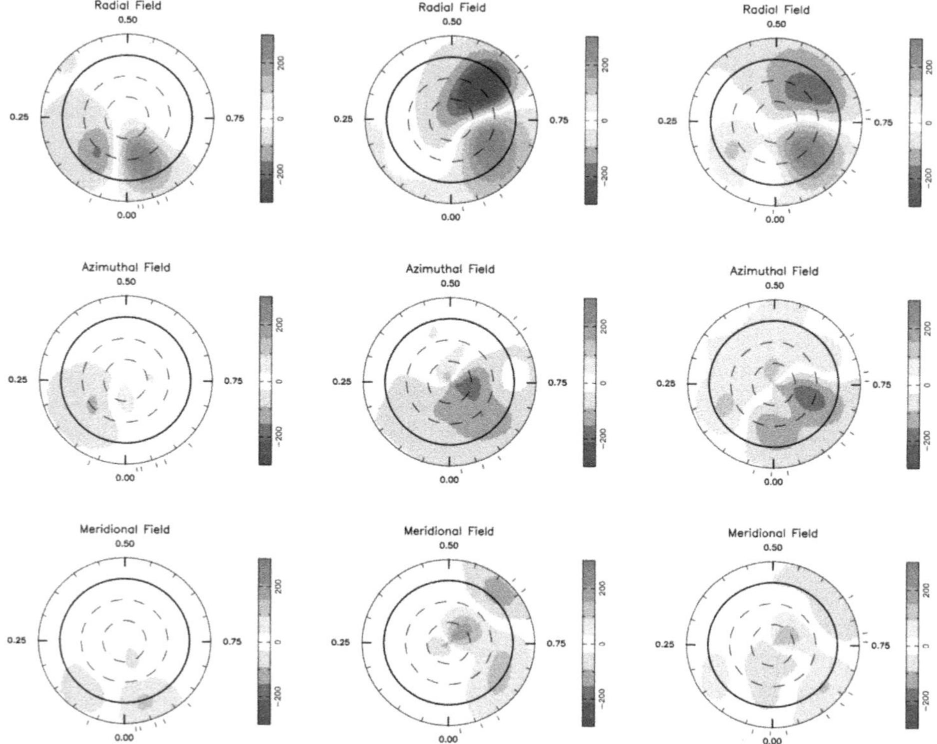

Figure 6.11. Maps of the radial, azimuthal, and meridional components of the uncancelled magnetic field on the flare star GJ 1156. Opposite polarities are shown with red and blue and the (incomplete) phase coverage of the observations are shown with the outside radial tics. The 3 epochs are 2007, 2008, 2009 from left to right. The vertical bars give the scale of the uncancelled field in Gauss. Credit: Reproduced from Morin et al. (2010). © 2010 The Authors, Monthly Notices of the Royal Astronomical Society © 2010 RAS.

numerous smaller bipolar regions. A comparison of total flux measured using Stokes I with uncancelled flux measured using Stokes V appears in Figure 6.13 from KoRev21. It shows that the largest total fluxes along with uncancelled fluxes are on low-mass stars with small Rossby numbers (rapid rotators). There also seems to be a set of low-mass stars with fairly low Rossby numbers that have less field in more complicated configurations. It is currently unclear why this difference exists.

Unsurprisingly, stars with axisymmetric configurations tend to have a larger percentage of their field uncancelled because those structures are generally larger in the same polarity. The fully convective boundary is at 0.3 solar masses, and none of the stars above that mass have less than 90% of their flux canceled in the ZDI maps. These results show that it is quite useful to do both types of magnetic field measurement at the same time for the same star. At the same time, it is clear that further precision and sophistication in the direct measurement of magnetic fields and their configurations is quite desirable in the future.

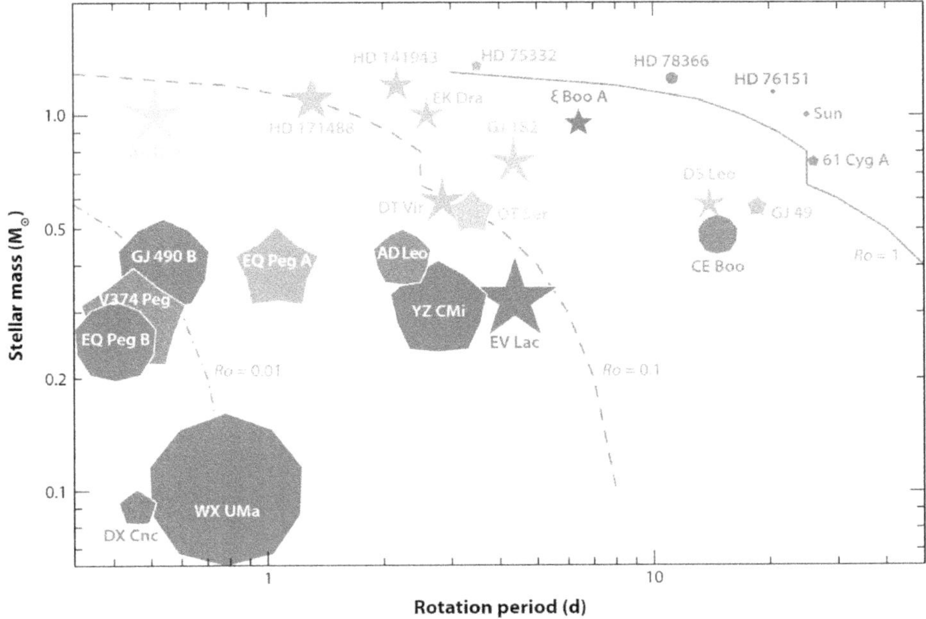

Figure 6.12. A summary of results on (uncancelled) magnetic flux magnitudes and field configurations inferred from ZDI measurements. The size of the symbol indicates the magnitude, the color indicates whether the field is dominantly toroidal (red) or poloidal (green, blue), and the shape indicates axisymmetry (roundish) or non-axisymmetry (pointy). Dividing lines in Rossby number are shown with lines. Credit: Reprinted from Donati et al. (2009), with the permission of AIP Publishing.

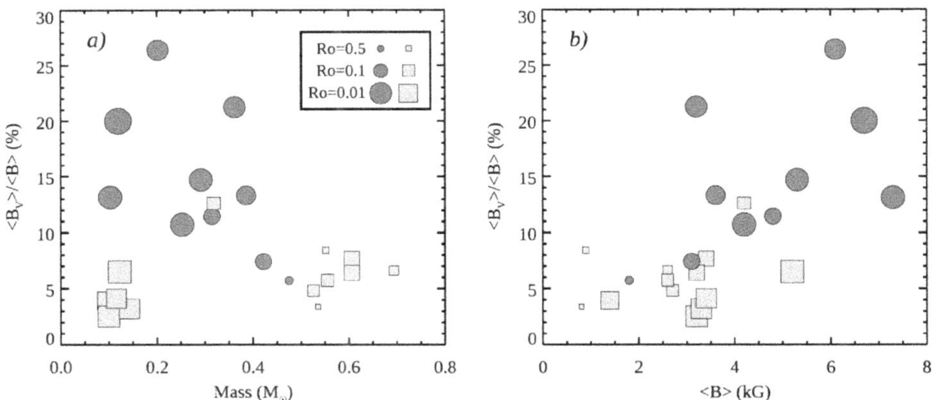

Figure 6.13. The ordinate shows the percentage of the uncancelled flux to the total flux for cases in which both are known. The left plot is the relation to stellar mass and the right plot is the relation to mean field strength. Symbols are larger for smaller Rossby numbers. Circles indicate simpler (dipolar, axisymmetric) cases and squares have more complicated ZDI configurations. Credit: Reprinted by permission from Springer Nature: Kochukhov (2021), © 2021, The Author.

References

Baliunas, S., Frick, P., Moss, D., et al. 2006, MNRAS, 365, 181

Baliunas, S. L., Donahue, R. A., Soon, W. H., et al. 1995, ApJ, 438, 269

Basri, G., Marcy, G. W., & Valenti, J. A. 1990, ApJ, 360, 650

Basri, G., Marcy, G. W., & Valenti, J. A. 1992, ApJ, 390, 622

Brandenburg, A., Mathur, S., & Metcalfe, T. S. 2017, ApJ, 845, 79

Browning, M. K. 2008, ApJ, 676, 1262

Browning, M. K., Miesch, M. S., Brun, A. S., & Toomre, J. 2006, ApJ, 648, L157

Chabrier, G., & Küker, M. 2006, A&A, 446, 1027

Charbonneau, P. 2010, LRSP, 7, 3

Dikpati, M., McIntosh, S. W., Chatterjee, S., et al. 2019, NatSR, 9, 2035

Domingo, V., Ortiz, A., Sanahuja, B., & Cabello, I. 2005, AdSpR, 35, 345

Donati, J. F., & Landstreet, J. D. 2009, ARA&A, 47, 333

Donati, J.-F., Morin, J., Delfosse, X., et al. 2009, in AIP Conf. Proc. 1094, Cool Stars, Stellar
   Systems, and the Sun (Melville, NY: AIP), 130

Donati, J. F., Semel, M., Carter, B. D., Rees, D. E., & Collier Cameron, A. 1997, MNRAS, 291,
   658

Durney, B. R., & Robinson, R. D. 1982, ApJ, 253, 290

Finley, A. J., & Matt, S. P. 2017, ApJ, 845, 46

Hall, J. C., Lockwood, G. W., & Skiff, B. A. 2007, AJ, 133, 862

Hathaway, D. H. 2015, LRSP, 12, 4

Hathaway, D. H., Nandy, D., Wilson, R. M., & Reichmann, E. J. 2003, ApJ, 589, 665

Hathaway, D. H., & Rightmire, L. 2010, Sci., 327, 1350

Isık, E., Solanki, S. K., Krivova, N. A., & Shapiro, A. I. 2018, A&A, 620, A177

Johns-Krull, C. M., & Valenti, J. A. 2000, in ASP Conf. Ser. 198, Stellar Clusters and
   Associations: Convection, Rotation, and Dynamos (San Francisco, CA: ASP), 371

Kochukhov, O. 2021, A&ARv, 29, 1

Kochukhov, O., & Lavail, A. 2017, ApJ, 835, L4

Küker, M., Rüdiger, G., & Schultz, M. 2001, A&A, 374, 301

Lubin, J. B., Rodriguez, J. E., Zhou, G., et al. 2017, ApJ, 844, 134

Marcy, G. W. 1984, ApJ, 276, 286

Montet, B. T., Tovar, G., & Foreman-Mackey, D. 2017, ApJ, 851, 116

Morin, J., Donati, J. F., Petit, P., et al. 2010, MNRAS, 407, 2269

Muirhead, P. S., Veyette, M. J., Newton, E. R., Theissen, C. A., & Mann, A. W. 2020, AJ, 159, 52

Namekata, K., Davenport, J. R. A., Morris, B. M., et al. 2020, ApJ, 891, 103

Netto, Y., & Valio, A. 2020, A&A, 635, A78

Reiners, A. 2012, LRSP, 9, 1

Reiners, A., & Basri, G. 2007, ApJ, 656, 1121

Reinhold, T., Cameron, R. H., & Gizon, L. 2017, A&A, 603, A52

Rueedi, I., Solanki, S. K., Mathys, G., & Saar, S. H. 1997, A&A, 318, 429

Saar, S. H., & Linsky, J. L. 1985, ApJ, 299, L47

Schrijver, C. J., Cote, J., Zwaan, C., & Saar, S. H. 1989, ApJ, 337, 964

Usoskin, I. G., Solanki, S. K., Krivova, N. A., et al. 2021, A&A, 649, A141

Valenti, J. A., Johns-Krull, C. M., & Piskunov, N. E. 2001, in ASP. Conf. Ser, 223, Cool Stars,
   Stellar Systems and the Sun, ed. R. J. Garcia Lopez, R. Rebolo, R. J. Garcia Lopez, &
   M. R. Zapaterio Osorio (San Francisco, CA: ASP), 1579

Valenti, J. A., Marcy, G. W., & Basri, G. 1995, ApJ, 439, 939
Vidotto, A. A., Jardine, M., Morin, J., et al. 2014, MNRAS, 438, 1162
Yadav, R. K., Christensen, U. R., Morin, J., et al. 2015, ApJ, 813, L31
Yadav, R. K., Christensen, U. R., Wolk, S. J., & Poppenhaeger, K. 2016, ApJ, 833, L28

An Introduction to Stellar Magnetic Activity

Gibor Basri

Chapter 7

Stellar Magnetism in Other Contexts

The book so far has concentrated on the observations, theory, and techniques related to activity and magnetic fields in the atmospheres of main sequence stars with outer convective envelopes. This final chapter starts with a trip to the bottom of the main sequence and into the realm of brown dwarfs. Classical stellar activity fades out along the way, although strong magnetic fields persist and non-thermal emission is sometimes seen in extremely cool objects.

We next take a brief tour of other types of stars with interesting magnetic fields, even if they don't always exhibit the sort of stellar activity presented earlier. Newly forming (T Tauri) stars have among the strongest magnetic fields on stars with convective envelopes, and early on these can interact with the accretion disk that is still building the star up. A stars were among the first on which fields were directly detected because they sometimes display very strong fields. They can produce fields via a dynamo in their convective nuclear cores that makes it to the surface or may retain some primordial field. OB stars don't produce magnetic fields in their envelopes but can retain strong primordial fields since they are massive and short-lived. O stars can also produce non-radiative heating in their outer atmospheres via radiatively-driven shocks.

The book ends with a brief introduction to a topic that has revitalized the entire field: the effect of stellar activity on exoplanets. Each of these topics could easily warrant a whole chapter. My choice to treat them more briefly here is subjective and does not mean they are less important than the topics I have chosen to emphasize. I have fully left out the subject of magnetic fields on stellar corpses including magnetic white dwarfs, pulsars, and magnetars. Magnetism has a pervasive presence in astrophysics.

7.1 The Bottom of the Main Sequence (and Below)

A puzzle in the behavior of magnetic activity (which will primarily be represented by $H\alpha$ emission in this section) in very low-mass stars is that the $H\alpha$ equivalent widths

doi:10.1088/2514-3433/ac2956ch7

are about 10 Å or less nearly all the way to the bottom of the main sequence. This can be seen in Figure 7.1 from Newton et al. (2017). As explained by Basri & Marcy (1995) the naive expectation is that Hα emission should not only be easier to detect (Figure 7.1 does show an increasing activity fraction) but also appear steadily stronger in cooler stars. By the latest M sub-types one might expect equivalent widths an order of magnitude larger than observed for the following reasons. The source function for Hα is complicated and non-local thermodynamic equilibrium (NLTE) as described in Section 3.3, but one might expect it to attain a relatively consistent value since chromospheric excitation of level 2 (independently of photospheric effective temperatures) is required to excite Hα at all. The source function will also be closer to thermalized in very cool stars because their surface gravities and thus atmospheric densities are greater. The equivalent width is really a contrast measure between the source function at line center compared with the source function at nearby continuum wavelengths. For solar-type stars that nearby continuum is truly the thermal photosphere, but for M stars it is hidden by a forest of molecular lines whose source function gets rapidly fainter as one moves into late-M and early-L spectral types due to their decreasing effective temperatures. The contrast between the relatively constant line core and the rapidly decreasing outer wings, and so the equivalent width, ought to grow rapidly but doesn't. One possible explanation is the increasing neutrality of photospheric plasma that could reduce magnetic dissipative heating as discussed below, but level 2 must still be populated. This hypothesis requires further research or another explanation must be found.

Beyond the question of equivalent widths, an activity luminosity ratio will also tend to rise as the bolometric luminosity decreases unless compensated by a greater decrease in the activity diagnostic. A presentation of $\log(L_{H\alpha}/L_{bol})$ for stars above and below the main sequence is shown in Figure 7.2 from Reiners & Basri (2008). The conversion of Hα equivalent widths into activity ratios with bolometric luminosity is a little tricky and different authors have used somewhat different prescriptions. A recent discussion of conversion factors is provided by Newton et al. (2017). Figure 7.2 shows

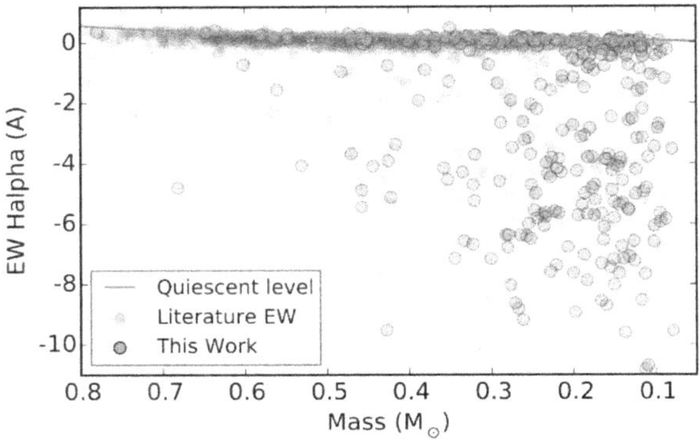

Figure 7.1. Hα equivalent widths for a large sample of M dwarfs. The spectral resolution was generally moderately high; negative numbers indicate emission. Credit: Reproduced from Newton et al. (2017). © 2017. The American Astronomical Society. All rights reserved.

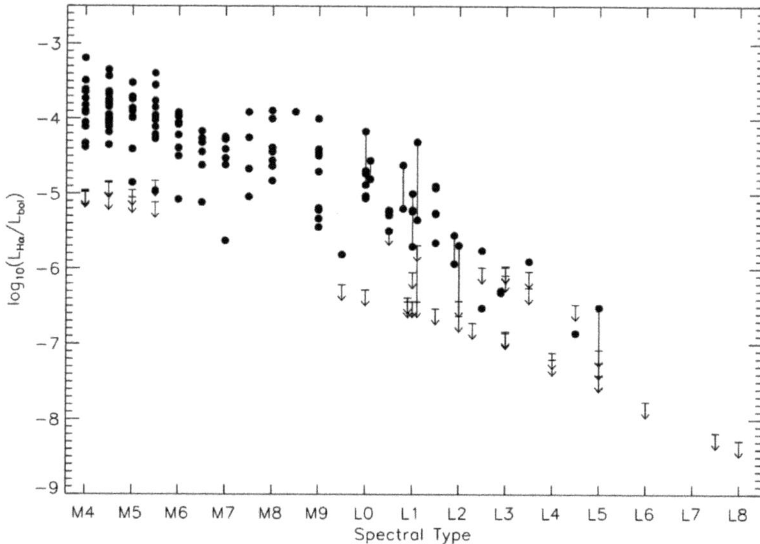

Figure 7.2. The behavior of $\log(L_{H\alpha}/L_{bol})$ with spectral type from high resolution spectra of main sequence stars. The filled circles are stars with $H\alpha$ emission and the down arrows indicate upper limits on lack of emission. Credit: Reproduced from Reiners & Basri (2008). © 2008. The American Astronomical Society. All rights reserved.

that this luminosity ratio decreases steadily with later spectral types. It also shows that the fraction of stars with $H\alpha$ emission is very high in the later M stars but then starts decreasing fairly rapidly and is quite low by mid-L spectral types. The bottom of the main sequence is somewhere around L3; cooler objects are all brown dwarfs and a few of the warmer objects are too (when they pass the lithium test). The luminosity ratio decreases with stellar mass (later spectral types) as well as age (determined kinematically). Because the numerator is measured with respect to the pseudo-continuum level at 656 nm (via equivalent width) that means a further decrease in the absolute energy that comes out in $H\alpha$ emission. In both cases, however, the flux at that wavelength decreases with effective temperature. If ultraviolet (UV) and X-rays behaved in the same fashion as $H\alpha$ (more observations are needed to establish this) the problem of close exoplanet habitable zone proximity to low-mass stellar activity would be reduced, but only for objects at or below the bottom of the main sequence.

The first hint that something completely new happens near the bottom of the main sequence was discovered in the M9.5+ dwarf BRI 0021-0214 by Basri & Marcy (1995). Observations with the newly commissioned Keck I telescope had just allowed the first high resolution spectroscopy of late-M dwarfs. The expected forest of molecular absorption lines was absent, as was any $H\alpha$ emission. I realized that the absence of distinct molecular lines is due to extreme Doppler broadening arising from a rotation period of a few hours. This rate of spin was unheard of for field M dwarfs (the paper found similar rotation velocities for very young stars but they are much larger). Later work has shown that rotation periods of hours are typical for brown dwarfs, reminiscent of the rotation periods of the giant planets in our solar system.

A summary of results for very low-mass objects was provided by Reiners & Basri (2008). Up to then rotation was largely measured through Doppler broadening, which for these objects usually had a lower detection limit of $v \sin(i) = 3$ km s^{-1}. A given velocity translates into different rotation periods for stars of different radii. Figure 7.3 shows that the early M stars observed are all at or below that detection limit; higher velocities begin to show up in the mid-M dwarfs. By late M all the stars have detected rotation and the median rotation velocity is steadily increasing despite their decreasing radii. Near and below the bottom of the main sequence there are no stars with broadening less than 20 km s^{-1}, so we are in the domain of very short periods since these are also the smallest objects. Reiners & Basri (2008) argued that this requires a mass-dependent braking law in which braking becomes decreasingly efficient at very low temperatures and provide a possible formulation for it. A caveat now apparent is that this sample does not contain the late-type slow rotators seen in Figure 5.9 found by Newton et al. (2017).

The even more startling surprise found in BRI 0021 by Basri & Marcy (1995) is the absence of any Hα emission. As has been clearly established in Chapter 5, one expects activity diagnostics to increase as the rotation period decreases and this star shows a hugely decreased period. Furthermore it was explained just above that one expects Hα emission to be more easily seen and stronger for cooler objects. Their measurement was quite sensitive to such emission and established its equivalent width was less than 0.2Å. Later the same star did exhibit Hα flare emission (Reid et al. 1999) but those authors estimated that it shows this less than 10% of the time and at a relatively low level. The strong implication is that the rotation–activity

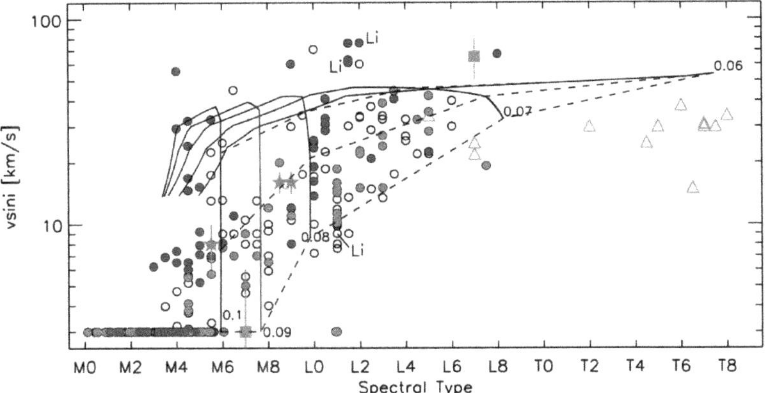

Figure 7.3. Measurements of the rotational broadening in very low-mass objects as of 2008. Blue dots represent kinematically young objects and red dots are older, while open circles did not have that information. Gray triangles cover the coolest objects from another survey and other symbols stand for special cases. The dashed lines are isochrones for 2, 5, and 10 Gyr, and the solid lines are braking models for objects of different masses. Credit: Reproduced from Reiners & Basri (2008). © 2008. The American Astronomical Society. All rights reserved.

connection that until then seemed to be an ironclad law breaks down at the bottom of the main sequence, heralding the death of normal stellar activity. Later observations confirmed this, as presented in Figure 7.2.

An explanation for this qualitative change was provided by Mohanty et al. (2002). As an example, for atmospheric temperatures of 2000 K and typical atmospheric densities there might be one ion for every billion neutrals. This is the more conservative estimate assuming backwarming by the dust that begins to form in such atmospheres; the ionization fraction drops by a few more orders of magnitude in clear atmospheres. The combination of very low ionization fraction and high density in these atmospheres results in very large resistivities and thus efficient field diffusion. While both ambipolar diffusion and Ohmic decay of currents due to ion–electron collisions occur, the primary diffusion effects are due to current decay through collisions of charged particles with neutrals. This resistivity is a strong function of both effective temperature and optical depth, increasing rapidly as either effective temperature or optical depth decreases. This implies that any magnetic field present is increasingly decoupled from atmospheric fluid motions as one moves from mid-M into the L spectral range. In late-M and L dwarfs, therefore, atmospheric motions are increasingly unable to force potential field configurations into highly dissipative (through reconnection) ones. That is, the magnitude of magnetic stresses generated by atmospheric motions becomes increasingly small in these objects.

Figure 7.4 from McLean et al. (2012) shows some of the effects of the fading out of classical stellar activity. It clearly shows that the usual relations between rotation and activity are broken in very cool low-mass objects for both Hα emission and X-rays (bottom and middle panels). It is not that activity has completely disappeared for these objects, although in cooler brown dwarfs it really does seem to be gone. Here we see that down to the bottom of the main sequence there is increased decoupling between rotation and activity with activity being suppressed compared to expectations and more scattered. Recall that for warmer objects, fast rotation (low Rossby numbers) resulted in saturation of the activity at the highest level (e.g. Figure 5.2 or figures 6 and 7 in Newton et al. 2017). Note the very compressed logarithmic ordinates in Figure 5.2; the suppression is up to two orders of magnitude. Although there are many fewer detections, the upper panel shows that radio luminosities are showing the opposite tendency and increasing with rotation velocity above the level of the higher-mass portion of the lower-mass group.

The radio data are useful in other ways for diagnosing what is going on at very low masses. For one thing, the Güdel–Benz relation (GBR; Section 4.2) is increasingly violated, as shown in Figure 7.5. That figure looks more definitive than it really is because there are a lot of radio upper limits which theoretically could actually belong on the GBR, but a number of those stars have shown flares that imply their quiescent values are above the relation. The faintest stars would not be detectable by current instruments if they lay on the GBR. On the other hand the detected low-mass stars lie an order of magnitude or two above the GBR. What is very useful about these detections is that they are clearly from non-thermal emission processes involving magnetic fields. Highly circularly polarized radio emission, sometimes pulsed, can be attributed to electron-cyclotron maser emission (Kao et al.

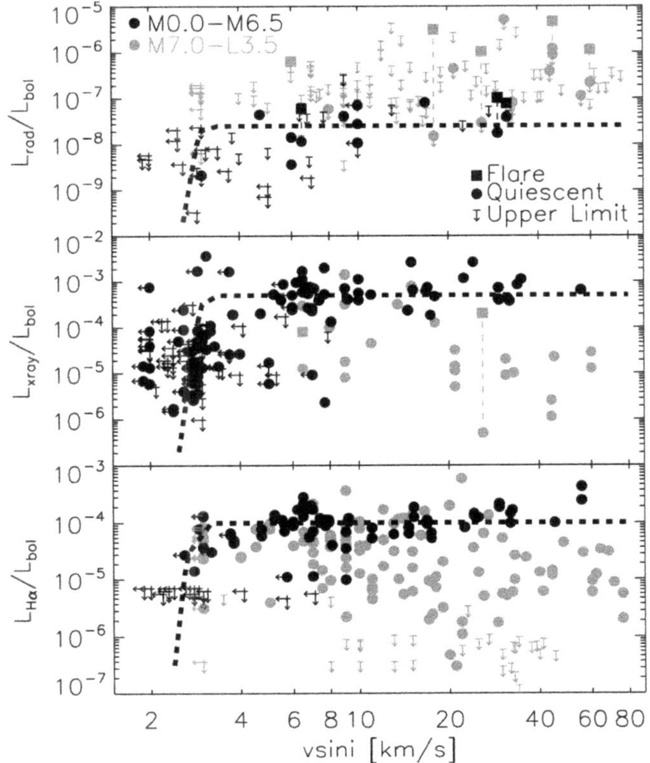

Figure 7.4. The behavior of three activity luminosity ratios near the bottom of the main sequence. Black dots are early to mid-M dwarfs and red dots are late-M to early-L dwarfs. Upper limits on both axes are indicated with arrows. In the middle and lower panels the lower-mass objects depart from the saturation behavior of the more massive (but still low-mass) ones and show increasingly low activity at higher rotation velocities. The top panel shows that radio luminosities have somewhat opposite behavior. Credit: Reproduced from McLean et al. (2012). © 2012. The American Astronomical Society. All rights reserved.

2016). This emission provides a direct way to measure the magnetic fields the electrons are circulating in and thus has proven that strong magnetic fields persist on brown dwarfs including some T and Y dwarfs.

The emission is similar to that seen from aurorae in our solar system, but they are unlikely to be produced like the Earth's. That involves the solar wind interacting with the planetary magnetic field. The stellar aurorae instead can be produced when an object with a strong field is rapidly rotating (as all these objects are) and corotation becomes difficult in the magnetosphere of the object. The exact mechanism is still not well understood nor is this explanation yet fully accepted. Currents so generated could flow down to the object and sometimes produce Hα emission or X-rays, though much weaker than expected from normal stellar activity. Resulting energy emission and opacity changes could even account for some of the photometric periodicity seen in these objects although it is most likely that "weather" in the form of jet streams, dust clouds, and clearer patches analogous

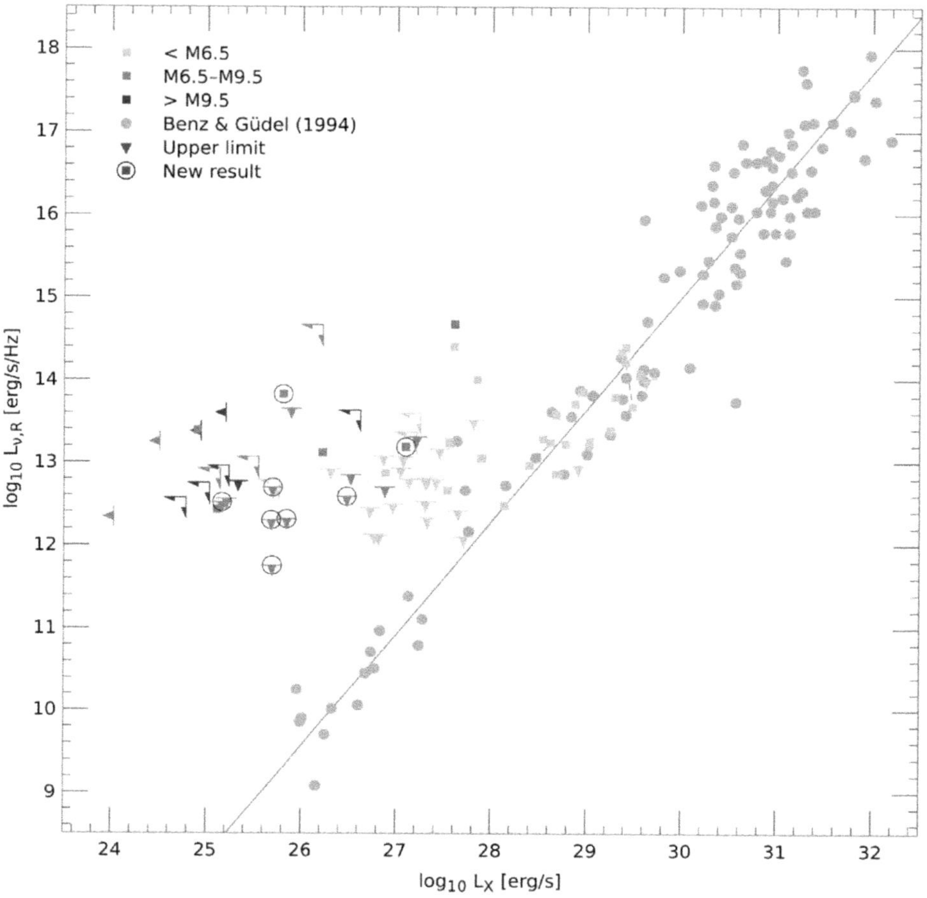

Figure 7.5. The relation between X-ray and radio emission for very low-mass objects compared to bigger stars. The gray dots represent the original data for the relation (cf Benz & Güdel 2010) with the dots at $L_X < 12$ coming from solar flares, then from flare stars up to $L_X < 14.5$ and mostly from active binaries above that. Detections do not include flares, however. Upper limits are shown with triangles and arrows. Credit: Adapted from McLean et al. (2012). © 2012. The American Astronomical Society. All rights reserved.

to what is seen in infrared images of Jupiter is mostly responsible. Searches for auroral brown dwarfs became much more successful when Hα emission was used to identify search targets. Of course there is a much larger population of similar objects from which such emission has not been seen. It would be good establish whether they simply have weaker fields or what is special about the emitting objects.

7.2 Pre-main Sequence Stars

Magnetic fields exist in the interstellar medium (ISM) at microgauss levels and they get concentrated as a molecular cloud core collapses to stellar scale by a factor of seven orders of magnitude or more. Furthermore dynamo processes are activated as the gathering star and disk spin faster and faster and become more turbulent. The newly forming star begins to shine, first due to gravitational contraction, then with a contribution from deuterium burning, then finally using steady hydrogen fusion to stop contracting at its main-sequence size. All but the most massive new stars are fully convective in the first part of their pre-main-sequence phase and all spin up as they contract. We have learned in earlier chapters that these conditions generate strong magnetic fields and it is now firmly established that is the case for these "T Tauri" stars (TTS).

The association of TTS with molecular clouds was an early clue that they could be stars in formation. Their spectra look similar to the solar chromosphere (although sometimes with much stronger Balmer lines) so there was a question whether this was an indication of very powerful stellar magnetic activity or might be due to accretion. There are clear indications of mass loss in some objects' Balmer line profiles and of inflows in others (sometimes both). Some TTS also exhibit spectral "veiling"—an apparent bluish continuum that causes the apparent strengths of absorption lines to be diluted (or in extreme cases eliminated). Finally it became clear with the advent of infrared astronomy that some TTS also exhibit IR excesses, yet their optical extinctions are incompatible with a spherical dust shell or cloud around them. An excellent review of the early historical development of the understanding of TTS can be found in section 2 of Bertout (2007); the whole volume containing that article has excellent reviews on TTS.

The 1980s saw great progress on these questions. Bipolar flows were discovered in the radio, sometimes associated with Herbig-Haro objects (shocked jet fronts in the ISM near TTS). Infrared, UV, and X-ray astronomy contributed a much more detailed understanding of the varied conditions on and around TTS. Spectral energy distributions (SEDs) covering the full wavelength range caused Shu et al. (1987) to propose three classes of very young objects: Class I are fully embedded in their birth clouds with mid- and far-IR SEDs and not optically visible (protostars), Class II are the classical TTS (CTTS) which still have strong IR and UV excesses but the star is optically visible, and Class III have nearly a normal stellar SED and strong stellar activity. Class III objects are also called weak-line TTS (WTTS) although their emission lines are only weak relative to the CTTS and actually quite strong relative to active stars. There is strong evidence from both semi-empirical modeling (Calvet et al. 1984) and observations (Finkenzeller & Basri 1987) that there are indeed very strong chromospheres on TTS. If sufficiently deep they can even produce Balmer and Paschen continuum emission that would explain the veiling, although they would tend to produce more emission cores in strong lines.

As the decade progressed it became increasingly clear that the disk paradigm was the correct explanation for many CTTS phenomena in tandem with strong stellar activity (for details and references see Section 2 of Bertout 2007). The IR observations made increasingly clear that there can be a lot of dust in the vicinity of a CTTS yet the star does not suffer the extinction it would cause if in front of the star. The bipolar outflows could be caused by a slow broad molecular disk wind and the fast collimated jets could be produced by magnetic fields near the star. Optical forbidden lines from jets showed a morphology that strongly suggested disks because the receding flow is often blocked from view but the approaching flow is visible. Bertout et al. (1988) showed that the IR excess could be from a disk and the UV excess could be from accretion from the disk onto the star.

Because this book is about stellar magnetic fields I will not discuss the rich topic of disks further except regarding their interaction with those fields. In that context Bertout et al. (1988) also presented evidence that the accretion onto DF Tau was modulated by the stellar rotation period and suggested that it was affected by the stellar magnetosphere. The luminosity and temperature of the accretion also suggests that it only covers a small fraction of the stellar surface. The rapid variability of the emission lines (hours and days as well as longer timescales) had been known for some time, but became a subject of more intensive observational campaigns with the higher spectral resolution that was becoming available (a review of this topic was provided by Basri 2007). These short durations suggest that the changes must be happening near the star and both inflow and outflow signatures can have this characteristic.

As mentioned earlier, Basri et al. (1992) established that very strong magnetic flux is present on WTTS; it was expected (and later confirmed) to be at least as strong on CTTS. Not long after Edwards et al. (1994) argued that the behavior of the inflow signatures (particularly in Hβ) suggested that accretion occurred as free-fall down stellar field lines. Shu et al. (1994) produced a detailed model that cemented the idea of magnetospheric accretion as well as explaining the fast collimated jets from CTTS as due to the interaction of the stellar field with the disk in an "X-wind" (Figure 7.6). This model presumes that the disk is interrupted by the stellar magnetic field at about the corotation radius (several stellar radii). Johns & Basri (1995) showed explicit evidence in a time series of Hα profiles from the CTTS SU Aur that both accretion and outflows take place on opposite sides of the star with the stellar periodicity as predicted by this model.

Attempts to check the model with actual measurements of magnetic field strengths and rotation periods have indicated that the real situation is more complicated than the simple model implies (Bouvier et al. 2007), but the model is qualitatively justified. Further discussion of the magnetic field measurements was given in Section 6.2. There have been many measurements of Zeeman broadening in the infrared (e.g., Johns-Krull & Valenti 2000) and many ZDI images of both CTTS (Donati et al. 2008) and WTTS (Carroll et al. 2012) have been made (cf Donati & Landstreet 2009). Many studies of multiple line profiles, sometimes with time variability of many TTS have been published (e.g., Alencar & Basri 2000; Edwards et al. 2006; Ardila et al. 2013). An example of several profiles from one star is given in Figure 7.7. The Hβ profile shows the accretion absorption at +200 km s^{-1} and the He I IR line shows outflowing absorption on the blueshifted side. The Ca II IR line shows the narrow chromospheric component at the stellar velocity and the broad magnetospheric component. The hydrogen lines are harder to interpret but clearly have components from off the star. There has been a great deal of modeling of line profiles and MHD simulations of magnetospheric accretion, nicely summarized by Hartmann et al. (2016). They reinforce that the environment around CTTS is very complicated and time variable because both the stellar

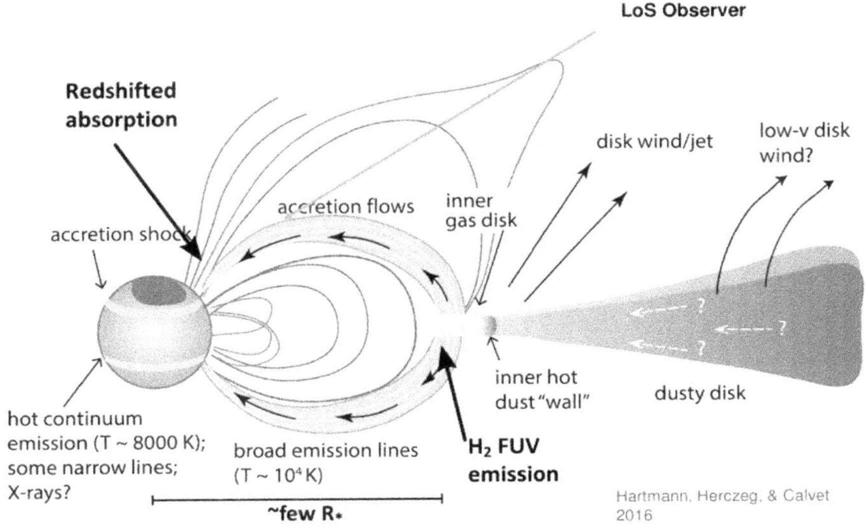

Figure 7.6. A schematic model of a CTTS undergoing magnetospheric accretion. In addition to the flow of material down stellar field lines that ends in an accretion shock near the stellar surface, there are outflow components driven by the field near corotation. This produces a rich set of spectral features in the continuum and lines over many wavelengths. Credit: Reprinted from Hartmann et al. (2016). © Annual Reviews.

magnetic field and the accretion disk undergo constant detailed changes. The explanations require a lot of observations and sophisticated modeling; this work will continue for some time into the future.

The main point from our perspective is that TTS have the strongest activity levels known. The Sun was a thousand times more active when the planets were forming than it is now. Flares on TTS can be up to ten thousand times more powerful than the Carrington event. The most powerful occur perhaps a hundred times more frequently than the strongest events on the Sun, and smaller flares occur still more often (many are stronger than current solar flares). All this magnetic activity is significant in the life of the star and possible planets. It is the primary cause of the loss of gas in the accretion disk (except in systems near OB stars); that happens over the first few million years. The angular momentum history of cool stars is influenced by whether they experience disk-locking with the stellar magnetosphere or not, then later by magnetic braking (Bouvier et al. 2014). It has been proposed that all the chondrules embedded in CaI-type meteors are produced through a process of making dust into molten drops near the termination of the disk by the stellar magnetosphere then launching them with the X-wind throughout the planetary system (Shu et al. 1996). This received support from the discovery of frozen rock droplets in a comet from the outer solar system where other melting processes are

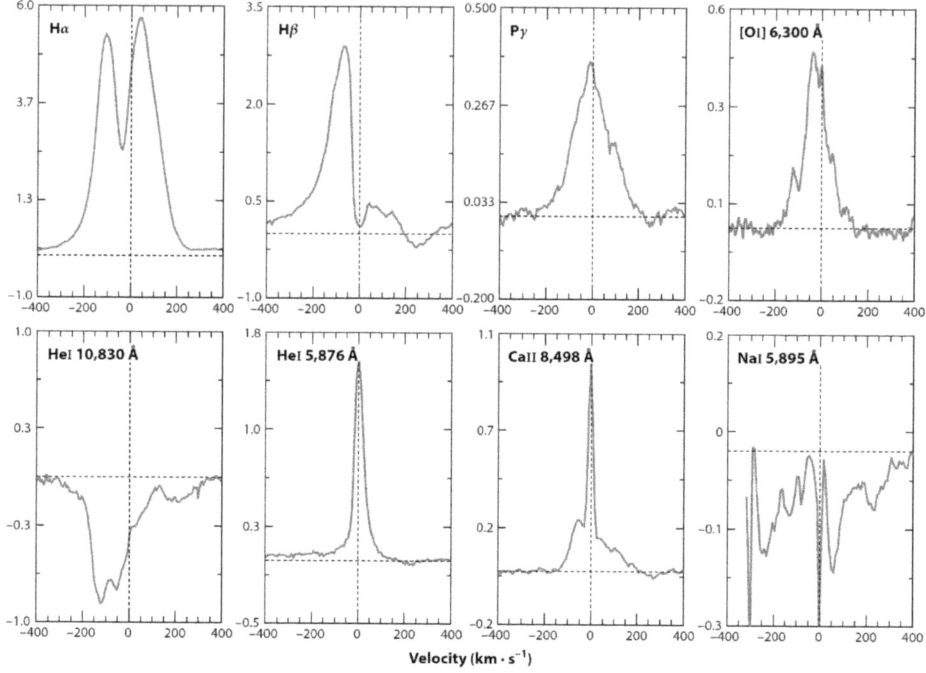

Figure 7.7. Line profiles from the CTTS DK Tau. The Hα and IR He I lines show typical broad emission and blueshifted wind absorption. Hβ and the sodium lines show redshifted infall signatures; the IR He I line shows both components. The Ca II triplet line shows the narrow chromospheric component (also seen in optical He I) and a broad magnetospheric component. The [O I] comes mostly from the jet far from the star. Credit: Reprinted from Hartmann et al. (2016). © Annual Reviews.

hard to come by. Finally the stellar activity is clearly responsible for influences on planetary atmospheres (Section 7.5).

7.3 Post-main Sequence Stars

As stars leave the main sequence they get brighter, larger, and cooler at the surface. A convective envelope develops if there wasn't one already present, and if there was the convective envelope deepens as the star evolves toward the red giant branch. Stars that had radiative envelopes on the main sequence do not undergo the magnetic braking that convective stars experience, so when they develop convective envelopes as subgiants they are rapidly rotating and so develop strong surface magnetic fields. It is also possible that even if the surface rotation has been braked, as the convective envelope deepens it begins to dredge up angular momentum from a more rapidly rotating core. There is some evidence for such a process in the "lithium-rich" stars. In the case of these moderately rapidly rotating giants (only about 1% of giants) a number of mechanisms have been proposed to explain their rotation and lithium abundances and this is an ongoing area of rapidly increasing information (e.g., Martell et al. 2021).

As the star expands one expects the rotation to slow simply on moment-of-inertia arguments and if it has developed fields magnetic braking should also play a role. It is therefore not surprising that red giants and supergiants (even more so) are slow rotators. They are slow enough that the sort of dynamo activity we have discussed in this book might not be effective, yet (weak, polarized) magnetic fields have been measured on some highly evolved stars. These include very bright well-known stars like Arcturus, Pollux, and Betelgeuse (Aurière et al. 2015). They tend to be stars with maximal convective velocities and there does appear to be a connection between rotation and activity, suggesting similar dynamo action to the main sequence stars.

Many subgiants and giants also show chromospheric emission lines like Ca II and Hα, often with wind features imposed onto the emission (Section 3.2). Some red giants have chromospheric fluxes that are consistent with being just basal (acoustic) with very weak to undetected magnetic fields, while others are more active. Rotation still seems to play a role, but now interactions with companions may explain some of the variety. A study of Ca II emission in giants and supergiants (Pasquini et al. 2000) shows the lowest gravity stars do not show emission but they are rather cool so perhaps calcium is not ionized. Recall that the width of the Ca II emission core is growing with luminosity up to that point (the Wilson–Bappu effect; Section 3.2). Gas hotter than chromospheric has been seen in stars with all but the lowest gravities (Dupree et al. 2005) and it appears that their outer atmospheres could be quite inhomogeneous (Section 4.4). Coronal plasma generally seems to become lower in temperature as the gravity gets lower and the winds get stronger (Section 4.2).

As stars become very luminous with very low gravities it is likely that the formation of dust and radiative driving of it plays an increasingly important role in the atmospheric and wind structures. Their atmospheres become increasingly extended and their convection cells become quite low in number and huge in size. The asymmetric structures seen in some planetary nebula have had magnetic fields invoked as one explanation. The role of magnetic fields in these phases of stellar evolution is probably not the same as in classic stellar activity.

In a different class and quite important in the study of stellar magnetic activity are close binary systems, some of which contain evolved components. In that case tidal locking can force the evolved star to continue to be a rapid rotator. These RS CVn systems can be extremely active and have provided much information on stellar activity in strong forms. Some of these systems are active and close enough that their magnetospheres can interact with each other (Figure 7.8) and there is evidence that some giant flares seen in those systems take place between the two stars (e.g., Trigilio et al. 1993). A more rare and extreme type of rapidly rotating giants are the FK Com stars. These very fast rotators are thought to be cases where the evolved component has actually swallowed its stellar companion and absorbed its angular momentum.

The RS CVn stars were first identified by their strong photometric modulation, presumed to be due to large starspots on one or both components. They also display high fluxes in all the usual emission lines and activity diagnostics that have been discussed in this book. They have particularly been the object of radio observations because of their high brightness, sometimes spatially resolved through interferometry. Mention of a number of the relevant historical references is in the introduction

Figure 7.8. A depiction of a close binary system with interacting magnetic fields. This sort of configuration can occur in RS CVn systems although this figure is of a potential field extrapolation by Holzwarth & Gregory based on ZDI studies of a close TTS system from Donati et al. (2011). Reproduced from https://discovery. dundee.ac.uk/en/publications/potential-magnetic-field-extrapolation-in-binary-star-systems with permission from V. Holzwarth and S. G. Gregory.

of Fox et al. (1994). These stars provided the first targets for Doppler imaging (Vogt & Penrod 1983) and the two-spot modeling discussed in Section 2.2. Some systems show a slow drift in the location of the light curve minima (slight period changes), which have been interpreted as differential rotation, dynamo waves, or other possibilities (e.g., Lindborg et al. 2013). Period drifting can also be simply due to random rearrangements of spots that alter the distribution asymmetry (Section 2.2), but in Doppler images it appears more due to longitudinal migration of the downward extensions of polar spots.

Fox et al. (1994) provide a good example of the kind of simultaneous multi-wavelength campaign that is needed to best understand these variable stars. They observed simultaneously at cm, microwave, UV, and X-ray wavelengths. Some flares were also observed during the campaign. The paper confirms the general correlations between activity diagnostics that were inferred in non-simultaneous observations of RS CVn systems that are also seen in other stellar activity. In some flares the X-rays and radio observations do not behave the same way; this is also seen in very low-mass stars (Section 7.1). The likely explanation in both cases is that some radio emission comes from coherent electron emission processes with high brightness temperatures that do not have to involve the large numbers of electrons needed to produce the observed X-ray bremsstrahlung emission. On the other hand the quiescent radio emission seen in many systems seems to be thermal gyrosynchro-tron radiation. X-ray observations of RS CVn systems provide among the best information because they are so luminous and there are a number of relatively nearby systems. They provided some of the first evidence that as one looks at more and more luminous coronae they contain two or more distinct temperature components or sometimes a broad distribution of temperatures. Significantly hotter plasma than in the solar corona (tens of millions of degrees) is increasingly bright with increasing activity (cf Dempsey et al. 1993). This is also seen in flare stars, but the RS CVn coronae can be even hotter and have greater emission measures at these temperatures. They also flare frequently, with temperatures that have been measured up to hundreds of millions of degrees.

A few cases are eclipsing systems which provides a further means of getting more spatially-resolved information. The most studied of these systems is AR Lacertae. Walter et al. (1983) made time-resolved X-ray observations of this system in and out of eclipse along with UV observations of chromospheric and transition region spectral lines. These observations were further analyzed by Walter et al. (1987) resulting in the map in Figure 7.9. It contains information about the location of coronal and chromospheric activity in the context of the binary system, including a case of how coronal structures are affected by interacting magnetospheres. The paper shows how time-resolved observations can provide a form of spatial imaging on rotating stars, but they have to be fairly contemporaneous because active stars are also quite time-variable. There continues to be speculation about whether the presence of another star affects where heating occurs or even the dynamo itself via tidal interactions, but these observations didn't really help with that.

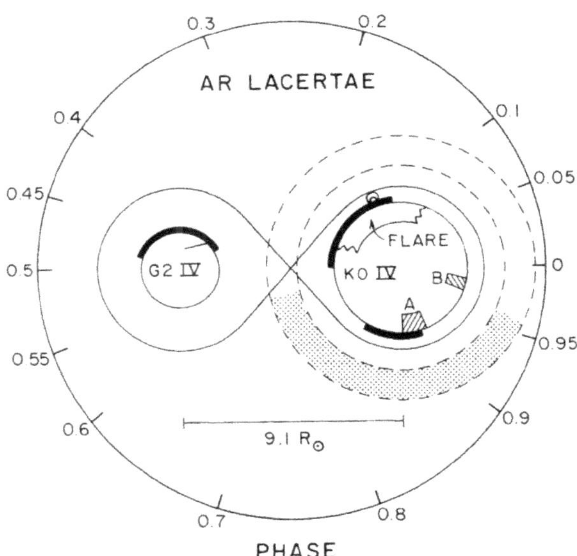

Figure 7.9. Surface and coronal structures derived on the RS CVn eclipsing system AR Lac They are inferred from both eclipse data in X-rays (Walter et al. 1983) and analysis (Walter et al. 1987) of time-resolved UV spectroscopy taken at the same time. The thick black lines indicate bright closed coronal loops, components A and B are regions of particularly bright chromosphere, and the stippled region is an extended coronal component of the evolving K star. The possible location of a flare based on X-rays is labeled, but its likely location constrained by Mg II is given by the small loops. The Roche lobes and orbital phases are also indicated. Credit: Reproduced from Walter et al. (1983). © 1983. The American Astronomical Society. All rights reserved.

7.4 Hot Main Sequence Stars

This section covers stars that are more massive than about 1.5 solar masses (spectral types early F and hotter) and therefore have radiative rather than convective envelopes on the main sequence. Although decades ago it was thought that magnetic fields might be absent on such stars because they don't have envelope dynamos, it was known even then that there is a small class of stars that show very strong stable configurations of magnetic field. Much of this section is a summary of the topics in section 5 of a very useful review by Donati & Landstreet (2009).

There are two good reasons why high-mass stars could possess magnetic fields. For the higher masses, the fields could be primordial (brought in from the ISM in the formation process) because those stars are short-lived enough. The Ohmic dissipation times are much longer in these hot highly-ionized envelopes so initial fields that survive the pre-main sequence convective phase can be retained in fairly stable configurations for much of the life of an A star (less than 1 Gyr). The other possibility arises because the nuclear-burning cores of high-mass stars are themselves convective and high-mass stars are generally rapid rotators (although it turns out many of the magnetic hot stars are comparatively slow rotators). The question then is whether fields generated by such an internal dynamo can manage to rise to the surface, and how stable they would be.

There is a small set of A and B stars designated Ap/Bp because their spectra show peculiarities in chemical composition, often with periodic changes. Another peculiarity is that they exhibit relatively narrow spectral lines indicating slow rotation compared to their spectral class. A few of them are slowly rotating enough to measure Zeeman broadening, and these usually have very strong fields (10–30 kG) with Babcock's star holding the record at 34 kG. The magnetic strengths in these stars do not seem to depend on their rotation rate as with solar-type stars. Only a few percent of high-mass stars show magnetic fields and their frequency increases with stellar mass. The field strength distribution is skewed toward low values, but the literature is skewed toward the strongest fields because they are easier to study and more interesting. The field spatial distributions seem to be stable in time. This is a point in favor of fossil fields since dynamo fields would likely be variable.

It is often the case that the magnetic and chemical signals are periodic at the rotation rate, indicating a non-axisymmetric distribution. The spectroscopic abundance variations also cause small photometric variations. The magnetic and spectral peculiarities are often in phase with each other, making it clear that the chemical and magnetic differences are related to each other. This is thought to be due to the fact that the field alters the competition between gravitational downward and radiative upward diffusion of elements with different atomic weights and opacities. Diffusion can be important because the shallow depths of these stars are very kinematically quiet given the absence of convection. The magnetic field is not relevant for neutral species but inhibits diffusion across field lines for ions. For detailed reviews of Ap/Bp stars see Landstreet (1992) and Linsky & Schöller (2015).

The rotations of the magnetic stars are often about an order of magnitude slower than other members of their spectral class. In extreme cases they can be three orders of magnitude slower (rotation periods of many years!). It is thought that most of the magnetic braking occurs in the pre-main-sequence phase when primordial fields are strongest and convective surface dynamos might operate for a bit. Obviously the braking is far greater for these very slow rotators for reasons that are not clear. Observations of Ap/Bp stars in open clusters with known ages have provided evidence that those between 2–5 solar masses continue to slow down by a factor of several during their main-sequence lifetimes. Interestingly, Ap stars with rotation periods less than a month seem to have their fields oriented perpendicular to their rotation axis. A map of one of these stars is shown in Figure 7.10. More slowly rotating Ap stars seem to have the axes more aligned.

There is also a set of Ap/Bp stars with much weaker fields (Aurière et al. 2007). These have primarily longitudinal fields detected with polarization. Their implied dipole components have a strength of at least 300 Gauss. The authors suggest there might be a critical field strength required to retain a large-scale field against differential rotation, which could also explain why higher-mass stars (more rapidly rotating) are less likely to show them. Very high-mass stars also begin to have increasingly fast and turbulent winds at their surface which could disrupt both the abundance patterns and the magnetic field.

The most massive stars have very short lives and so little difficulty in retaining primordial fields. It can be difficult to detect them on the usually very rapidly rotating O and B stars, but fields have been seen on a few O stars that are unusually slow rotators (or perhaps seen pole-on). The most extreme example is NGC 1624-2,

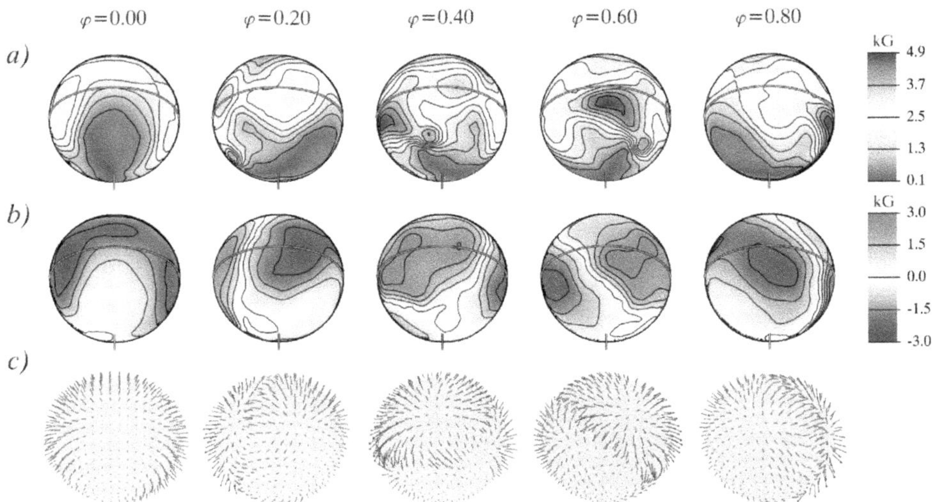

Figure 7.10. A ZDI map at five phases of the B9p bright star α^2 CVn ($P_{\rm rot} = 5$ days). All four Stokes components in a number of spectral lines from various heavy elements were utilized. The field map shows an asymmetric dipole-like configuration with poles at low latitudes, and other complex magnetic structures that are reproduced in abundance maps. The top row is the field modulus, the central row the radial field, and the bottom row the field orientation. Credit: Reproduced from Silvester et al. (2014). © 2014 The Authors, Monthly Notices of the Royal Astronomical Society © 2014 RAS.

which has the spectral designation O7f?p; other stars with similar designations tend to be magnetic as well. A field of around 20 kG has been detected on this star, which has a rotation period of 158 days and displays a strong wind and abundance anomalies. Its Hα emission goes through a cycle during the rotation that suggests we are looking at a single magnetic hemisphere with the pole pointing more toward and away from us as it rotates. Its ultraviolet emission lines (like C IV and Si IV previously discussed in the context of transition regions on solar-type stars) also show significant time variability, thought to be due to a dynamic magnetosphere that channels the wind along loops from both hemispheres that collide near the magnetic equator causing plasma to partially fall back onto the star. Another similar example is HD 191612 (Of?p) whose Hα emission line undergoes a periodic cycling between low and high states during its rotation period of 583 days. Sundqvist et al. (2012) did detailed modeling of its profile variations and explained them with a detailed model using 3-D MHD simulations and NLTE radiative transfer. Figure 7.11 shows some aspects of this model.

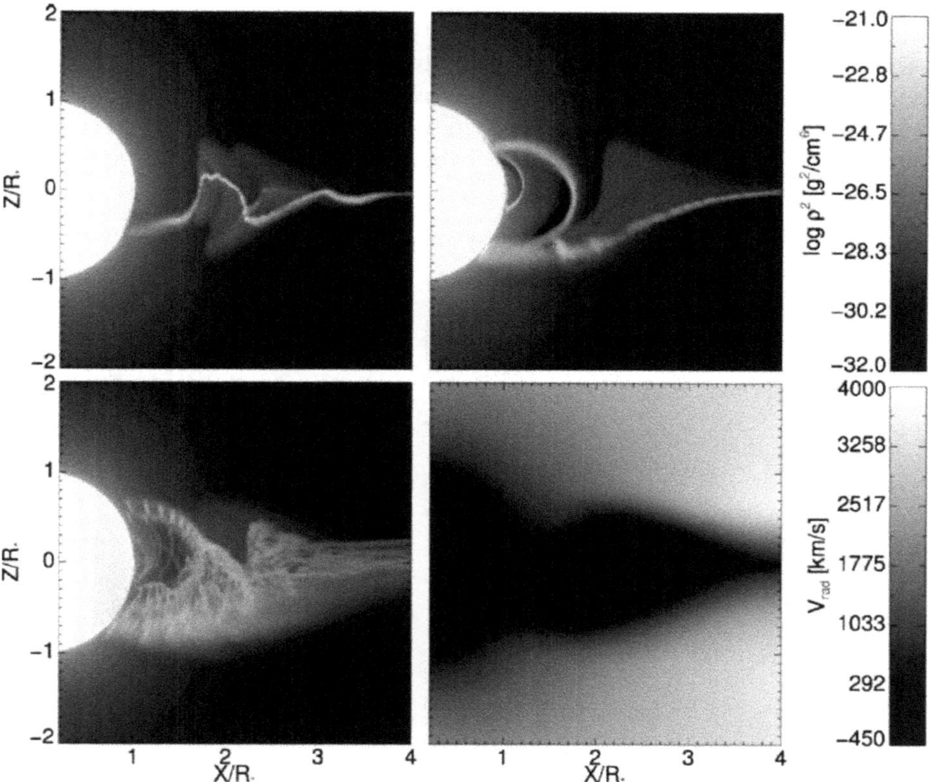

Figure 7.11. An MHD model of the dynamic magnetosphere of HD 191612. The upper panels show two different states in its dynamical evolution. The lower panels represent the density (left) and radial velocity (right) for the upper right case. Credit: Reproduced from Sundqvist et al. (2012). © 2012 The Authors, Monthly Notices of the Royal Astronomical Society © 2012 RAS.

Of course, when on the pre-main sequence high-mass stars are even more prone to show magnetic fields. The cooler of these are called "Herbig Ae/Be stars" because they show Hα emission and are associated with star-forming regions. As with TTS the lines are time-variable. There is another class of emission-line stars, the Be stars, but those are on the main sequence and their emission is not necessarily due to magnetic fields. The Herbig Ae/Be stars often have accretion disks around them, whereas the Be stars can actually have excretion disks caused by their very rapid rotation and oblate geometry. A well-known example of the former is AB Aur which has been monitored in optical and UV spectra for decades. Many of the emission lines show signs of rotational modulation suggesting control by magnetic fields (Catala et al. 1999). There is still confusion about the entangled roles of the fields, the stellar wind, and disk accretion in these complex objects. More recent studies with spectropolarimeters have shown that the Herbig Ae/Be stars seem to be the precursors to the Ap/Bp stars in terms of their magnetic field strengths and configurations (Alecian et al. 2008). Linsky & Schöller (2015) provide another review of these stars.

O and early B stars show strong soft X-ray emission (Güdel & Nazé 2009). It is largely in the form of line emission with a weaker component from bremsstrahlung, and appear to be mostly thermal. The X-ray luminosity is in the neighborhood of $10^{-7}L_{\mathrm{bol}}$ so it seems to be related to the strong fast radiatively-driven winds from these massive stars. The winds are expected to be somewhat unstable and subject to radiative instabilities in which slightly more optically thick clumps will experience relatively greater acceleration and collide with slower clumps ahead of them. Güdel & Nazé (2009) give a detailed discussion of the still not fully settled debate on the details of the X-ray production mechanisms. Coronal models have been tried but fail on several counts although models in which magnetic fields play some role have not totally been ruled out. In the context of this book it suffices to note that while very massive stars are absolutely luminous in X-rays, they probably are not caused by stellar magnetic activity.

7.5 Effects of Stellar Activity on Exoplanets

We know from the planets in our own solar system that stellar magnetic activity can have profound effects on planetary atmospheres. Of the four terrestrial planets, two of them have atmospheres that are quite different than they would be if the Sun had no magnetic field. The MAVEN spacecraft has made *in situ* measurements confirming that the solar wind is still slowly stripping atmospheric material away from Mars, and theory suggests that has been a major reason that Mars' atmosphere is quite a bit thinner now than it was on early Mars. Venus is the hottest planet because of its extreme greenhouse atmosphere due to 90 bars of carbon dioxide. It likely started off with a similar composition to the Earth, but its water remained in vapor form and was broken apart by the solar UV radiation, which we know was much stronger from the early Sun. The hydrogen was then ionized and stripped off by solar high-energy radiation and the solar wind. This prevented Venus' oceans from ever condensing and dissolving the carbon dioxide into rocky forms as happened on the Earth.

As a result of the Kepler mission we now also know that Earth-sized planets are very common and can be found anywhere in the inner planetary systems that were detectable by the transit method. In particular it seems that smaller stars tend to form smaller planets, so combined with the fact that smaller stars are more common it is likely that most Earth-sized planets in habitable zones are orbiting red dwarf stars. Because these stars have much lower bolometric luminosities than the Sun their habitable zones are much closer to them. On the other hand their stellar activity tends to have surface fluxes that are at least equal to that of the Sun and young red dwarfs have activity levels that far exceed the Sun. The time it takes for red dwarfs to become "quiet" (UV fluxes declined by more than an order of magnitude) depends on their mass, and ranges from 2–3 Gyr for star with half a solar mass to the age of the galaxy for the least massive stars (Section 5.3). It is important to understand the full spectrum of high-energy outputs from these stars. Many studies have been conducted in the last decade, including an empirical compilation by France et al. (2013) and semi-empirical modeling to reach the EUV by Linsky et al. (2014). An updated study of the decline in activity on early M stars (the most common) was given by Loyd et al. (2021).

The proximity of their habitable zones to red dwarfs means that the flux of high-energy radiation and particles steadily hitting their planets is greater by at least several hundred times (sometimes far worse) than what the Earth currently receives. Compounding that problem is the size and number of flares on red dwarfs. As mentioned earlier both the energies and frequencies of these flares can each be a few orders of magnitude greater than the current Sun produces. Each of these massive explosions can subject the planet to a pulse of high-energy radiation and particles that could substantially destroy an ozone layer like the Earth's (Tilley et al. 2019). Sometimes they cause the stellar luminosity itself to increase by several times for up to a few hours. Because we have far fewer X-ray and UV observations than optical observations, investigators have worked on converting the latter into the former (e.g., Youngblood et al. 2017). These confirm the severity of the problem. On the other hand, it should also be pointed out that these planets are so close to their star that they become tidally locked. Once that happens half the planet is no longer subjected to stellar emissions of any sort. Water could then be stored in a permanently frozen state on the dark side, perhaps melted near the interface with rock.

The result is that the planets of interest around red dwarfs are subjected to at least the fluxes of high-energy radiation that the Earth or Venus were during their earliest days but for far longer. The problem is compounded by the fact that the pre-main-sequence phase for red dwarfs can be up to 20 times longer than it was for the Sun. That means that planets that will end up in the habitable zone when the star is on the main sequence are much hotter during this time because pre-main sequence stars are more luminous (by several times) than they will settle down to. There is no chance that oceans could condense on habitable zone planets during this time, which can last well after the time we think it takes to form terrestrial planets. Of course, the stars are at their most active during this time as well, so the fluxes hitting these planets are far higher than our terrestrial planets ever experienced. This raises the question of whether Venus is a better model than the Earth for these inner terrestrial planets in the habitable zones of red dwarfs, or even worse, whether these planets can retain light elements in their atmospheres in any substantial amount. All these problems become rapidly less severe as one moves to K stars (Richey-Yowell et al. 2019) not only because the activity levels are lower and decline more rapidly but also because the habitable zones move further from the stars.

The loss of exoplanet atmospheres due to stellar activity is well established. In instances when a transiting planet with a thick atmosphere is located close to a star we can sometimes directly see it happening. The loss of hydrogen shows up as a much longer transit signature in Lyα than caused by the planet itself. The hydrogen essentially produces a "comet tail" behind the planet in its orbit that can cover much of the stellar disk. The first instance of this was observed from the hot Jupiter HD 209458b by Vidal-Madjar et al. (2004) using the Hubble Space Telescope. It also showed evaporation signatures in carbon and oxygen. Since then there have been a number of observations of other exoplanets losing their atmospheres; a recent example is Lavie et al. (2017). These observations confirm that stellar activity can strip substantial amounts of hydrogen and even heavier species like oxygen from exoplanets in amounts that depend on the distance between star and planet and the evolving levels of stellar activity. They cause the mass of gas giant planets to decrease a bit over time. It is a greater concern for smaller planets with thinner atmospheres (like the Earth) which could lose most of certain important elements from their atmospheres. Venus managed to retain a thick atmosphere but lost its hydrogen; planets close to red dwarfs could lose much more.

Another striking piece of evidence that this is an important process for close-in exoplanets is the planet radius gap observed between mini-Neptunes and super-Earths (Fulton & Petigura 2018). This gap occurs in the exoplanet size distribution at around 1.75 Earth-radii for close-in planets. The idea is that mini-Neptunes near that size have had their thick but low-mass hydrogen envelopes significantly depleted by photo-evaporation due to stellar activity, ending up looking more like rocky

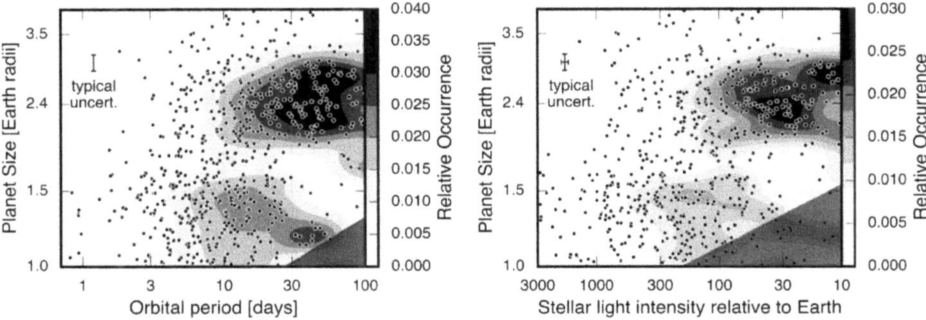

Figure 7.12. The two-dimensional distribution of exoplanet sizes as a function of orbital period and insolation. In both cases a clear gap at around a planet radius of 1.7 Earths is visible that separates mini-Neptunes (gas envelopes) from super-Earths (rocky). The super-earths are also more numerous very close to the star. Credit: Adapted from Fulton & Petigura (2018). © 2018. The American Astronomical Society. All rights reserved.

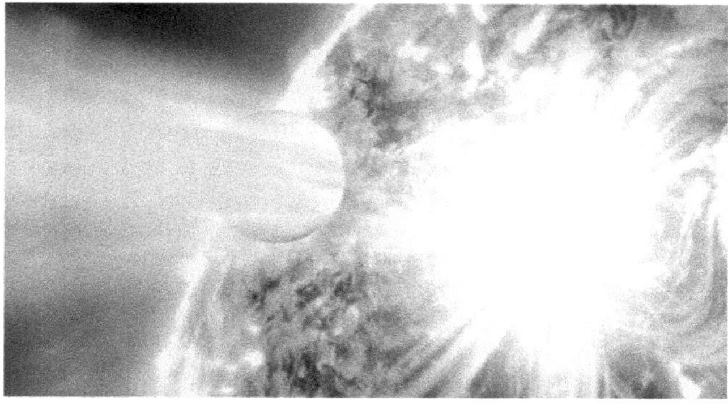

Figure 7.13. Artist's conception of a planet suffering the effects of stellar activity. Credit: Reprinted with permission from NASA.

super-Earths. The authors consider other explanations for this very clear gap (Figure 7.12) but show that photo-evaporation is the best explanation. The topic of how stellar activity affects exoplanet atmospheres (Figure 7.13) has only become really active in the past decade and is expected to be a major topic of research in the next decade. As we move forward in the search for life on other planets it is now appreciated that we must also understand the evolution and effects of magnetic activity on stars. For an extensive review of stellar activity in the context of exoplanets, consult the book by Linsky (2019).

References

Alecian, E., Catala, C., Wade, G. A., et al. 2008, MNRAS, 385, 391

Alencar, S. H. P., & Basri, G. 2000, AJ, 119, 1881

Ardila, D. R., Herczeg, G. J., Gregory, S. G., et al. 2013, ApJS, 207, 1

Aurière, M., Konstantinova-Antova, R., Charbonnel, C., et al. 2015, A&A, 574, A90

Aurière, M., Wade, G. A., Silvester, J., et al. 2007, A&A, 475, 1053

Basri, G. 2007, in IAU Symp. 243, Star-Disk Interaction in Young Stars (Cambridge: Cambridge Univ. Press), 13

Basri, G., & Marcy, G. W. 1995, AJ, 109, 762

Basri, G., Marcy, G. W., & Valenti, J. A. 1992, ApJ, 390, 622

Benz, A. O., & Güdel, M. 2010, ARA&A, 48, 241

Bertout, C. 2007, in IAU Symp. 243, Star-Disk Interaction in Young Star (Cambridge: Cambridge Univ. Press), 1

Bertout, C., Basri, G., & Bouvier, J. 1988, ApJ, 330, 350

Bouvier, J., Alencar, S. H. P., Harries, T. J., Johns-Krull, C. M., & Romanova, M. M. 2007, Protostars and Planets V, ed. B. Reipurth, D. Jewitt, & K. Keil (Tucson, AZ: Univ. Arizona Press), 479

Bouvier, J., Matt, S. P., Mohanty, S., et al. 2014, Protostars and Planets VI, ed. H. Beuther, R. S. Klessen, C. P. Dullemond, & T. Henning (Tucson, AZ: Univ. Arizona Press), 433

Calvet, N., Basri, G., & Kuhi, L. V. 1984, ApJ, 277, 725

Carroll, T. A., Strassmeier, K. G., Rice, J. B., & Künstler, A. 2012, A&A, 548, A95

Catala, C., Donati, J. F., Böhm, T., et al. 1999, A&A, 345, 884

Dempsey, R. C., Linsky, J. L., Schmitt, J. H. M. M., & Fleming, T. A. 1993, ApJ, 413, 333

Donati, J. F., Gregory, S. G., Montmerle, T., et al. 2011, MNRAS, 417, 1747

Donati, J. F., Jardine, M. M., Gregory, S. G., et al. 2008, MNRAS, 386, 1234

Donati, J. F., & Landstreet, J. D. 2009, ARA&A, 47, 333

Dupree, A. K., Lobel, A., Young, P. R., Ake, T. B., Linsky, J. L., & Redfield, S. 2005, ApJ, 622, 629

Edwards, S., Fischer, W., Hillenbrand, L., & Kwan, J. 2006, ApJ, 646, 319

Edwards, S., Hartigan, P., Ghandour, L., & Andrulis, C. 1994, AJ, 108, 1056

Finkenzeller, U., & Basri, G. 1987, ApJ, 318, 823

Fox, D. C., Linsky, J. L., Veale, A., et al. 1994, A&A, 284, 91

France, K., Froning, C. S., Linsky, J. L., et al. 2013, ApJ, 763, 149

Fulton, B. J., & Petigura, E. A. 2018, AJ, 156, 264

Güdel, M., & Nazé, Y. 2009, A&ARv, 17, 309

Hartmann, L., Herczeg, G., & Calvet, N. 2016, ARA&A, 54, 135

Johns, C. M., & Basri, G. 1995, ApJ, 449, 341

Johns-Krull, C. M., & Valenti, J. A. 2000, in ASP Conf. Ser. 198, Stellar Clusters and Associations: Convection, Rotation, and Dynamos, ed. R. Pallavicini, G. Micela, & S. Sciortino (San Francisco, CA: ASP), 371

Kao, M. M., Hallinan, G., Pineda, J. S., et al. 2016, ApJ, 818, 24

Landstreet, J. D. 1992, A&ARv, 4, 35

Lavie, B., Ehrenreich, D., Bourrier, V., et al. 2017, A&A, 605, L7

Lindborg, M., Mantere, M. J., Olspert, N., et al. 2013, A&A, 559, A97

Linsky, J. 2019, Host Stars and their Effects on Exoplanet Atmospheres, Vol. 955 (Berlin: Springer)

Linsky, J. L., Fontenla, J., & France, K. 2014, ApJ, 780, 61

Linsky, J. L., & Schöller, M. 2015, SSRv, 191, 27

Loyd, R. O. P., Shkolnik, E. L., Schneider, A. C., et al. 2021, ApJ, 907, 91

Martell, S. L., Simpson, J. D., Balasubramaniam, A. G., et al. 2021, MNRAS, 505, 5340

McLean, M., Berger, E., & Reiners, A. 2012, ApJ, 746, 23

Mohanty, S., Basri, G., Shu, F., Allard, F., & Chabrier, G. 2002, ApJ, 571, 469

Newton, E. R., Irwin, J., Charbonneau, D., et al. 2017, ApJ, 834, 85

Pasquini, L., de Medeiros, J. R., & Girardi, L. 2000, A&A, 361, 1011

Reid, I. N., Kirkpatrick, J. D., Gizis, J. E., & Liebert, J. 1999, ApJ, 527, L105

Reiners, A., & Basri, G. 2008, ApJ, 684, 1390

Richey-Yowell, T., Shkolnik, E. L., Schneider, A. C., et al. 2019, ApJ, 872, 17

Shu, F. H., Adams, F. C., & Lizano, S. 1987, ARA&A, 25, 23

Shu, F. H., Shang, H., & Lee, T. 1996, Sci, 271, 1545

Shu, F., Najita, J., Ostriker, E., et al. 1994, ApJ, 429, 781

Silvester, J., Kochukhov, O., & Wade, G. A. 2014, MNRAS, 440, 182

Sundqvist, J. O., ud-Doula, A., Owocki, S. P., et al. 2012, MNRAS, 423, L21

Tilley, M. A., Segura, A., Meadows, V., Hawley, S., & Davenport, J. 2019, AsBio, 19, 64

Trigilio, C., Umana, G., & Migenes, V. 1993, MNRAS, 260, 903

Vidal-Madjar, A., Désert, J. M., Lecavelier des Etangs, A., et al. 2004, ApJ, 604, L69

Vogt, S. S., & Penrod, G. D. 1983, PASP, 95, 565

Walter, F. M., Gibson, D. M., & Basri, G. S. 1983, ApJ, 267, 665

Walter, F. M., Neff, J. E., Gibson, D. M., et al. 1987, A&A, 186, 241

Youngblood, A., France, K., Loyd, R. O. P., et al. 2017, ApJ, 843, 31

Appendix A

Basic Concepts of Radiative Transfer

Radiative transfer theory provides an explanation of what happens to a radiation field as it moves through a star and photons interact with matter. It underpins essentially all observations of stellar atmospheres since we really only receive information on them through the radiation that leaves the star. Many students find it somewhat heavy going so I have tried to distill it to the minimum necessary and concentrated on the development of physical intuition more than mathematical rigor. The primary need to know what is here arises if one wants to understand how we derive physical information from spectroscopic line profiles. It is possible to understand a great deal of the content in the body of this book without knowing much about those details but if you want to understand spectral diagnosis this material is recommended.

The basic expectation from stellar structure theory is that a star is hottest at its core where the pressure is highest due to the weight of the full extent of the overlying stellar mass. Because the star is embedded in very cold empty space, thermodynamics dictates that this heat flows toward the surface and finally is radiated into the vacuum. We are only concerned here with the atmosphere of the star, so this condition leads to the expectation that the atmosphere should have a temperature gradient that decreases outward until it becomes thin enough to lose most of the radiation. The bottom of the atmosphere is taken as the place where significant amounts of radiation begin to escape into space: we call this the photosphere. It is a bit arbitrary where to draw this line and it also depends on the photon energy.

To provide a little more physical insight I review the concepts of "optical depth" and "source function." This book is not a treatise on radiative transfer so to go beyond the most basic level one should consult such a book. One of my favorites (especially while free and online) is "Radiative Transfer in Stellar Atmospheres" by R. J. Rutten (Rutten 2003; hereafter simply "Rutten") https://webspace.science.uu. nl/rutte101/rrweb/rjr-pubs/2003rtsa.book.....R.pdf. The basic equation of radiative transfer (Rutten section 2.1.2) describes the change of specific intensity I_ν along a path. Specific intensity can be conceptualized as a laser beam of photons in a very well-defined direction with no opening angle.

doi:10.1088/2514-3433/ac2956ch8   A-1

The specific intensity (beam) will change along a path length ds through material by addition of photons from its emissivity j_ν and by subtraction of photons by "extinction" α_ν. The extinction can be expressed as an opacity (blocking cross-section) per unit length. Opacity and emissivity can both be caused by either scattering and absorption. We will discuss the differences between them in more detail later, but at a basic level photons can be caused to change direction by scattering (either into or out of the beam) or they can be produced or absorbed in the beam by energy exchange with matter.

The most basic equation of radiative transfer expresses this simply as

$$dI_\nu(s) = j_\nu(s)ds - \alpha_\nu(s)I_\nu(s)ds. \tag{A.1}$$

Hereafter the path variable s will generally not be included since most equations will be local (refer to a particular value of s) but I will indicate when we are talking about the surface. If we now define the source function as $S_\nu \equiv j_\nu/\alpha_\nu$ and the optical depth as $d\tau_\nu(s) \equiv \alpha_\nu(s)ds$ then Equation (A.1) can be re-written after dividing by $\alpha_\nu(s)ds$ as

$$dI_\nu/d\tau_\nu = S_\nu - I_\nu. \tag{A.2}$$

The emergent intensity (which is all that can be observed) can be defined in terms of the vertical optical depth and the angle at which one looks into the star. The angle is important because when looking through a slant path, the total amount of matter that must be traversed to get to a given vertical optical depth will be greater the greater the slant angle. The cosine of the slant angle is designated μ; it is unity when looking straight into the star. The integral form of Equation (A.2) at the top of the atmosphere then becomes (integrating over optical depth):

$$I_\nu^0(\mu) = \left[\int_0^\infty S_\nu(t_\nu) \exp(-t_\nu/\mu)(dt_\nu/\mu)\right]. \tag{A.3}$$

Essentially all of observational stellar radiative transfer is contained in Equation (A.3). Taking the first terms in a Taylor expansion (the Eddington–Barbier approximation; Rutten section 2.2.1) this equation tells us that in an optically thick atmosphere, **the emergent intensity will be roughly the source function at optical depth unity**. It is precisely true if the source function is a linear function of optical depth. This deceptively simple concept will serve us well throughout the book; always return to it to understand what is observed.

Of course, for all stars except the Sun we cannot make observations at different values of μ. The quantity that is measured from the whole visible surface of the star is defined as the flux: $F_\nu^0 \equiv \int_0^1 \mu I_\nu^0(\mu)d\mu$, and the flux observed at the Earth is that quantity reduced by the angular diameter of the star for the observer.

The subtleties and physics that allows us to infer local physical conditions from spectroscopy come from the ways that matter produces optical depth and the source function as a function of frequency and position. They both depend on the populations of atomic and molecular energy levels. The source function also depends on whether photons are thermalized (exchanged with the local material energy pool) or scattered when they interact with matter. The material properties of emissivity and opacity depend in turn on physical variables like composition, temperature, density, and radiation field. For spectral lines there is further dependence on atmospheric quantities like bulk velocities, turbulence, and magnetic fields.

Even a quantity like temperature can be deceptive. The basic assumption is usually local thermodynamic equilibrium (LTE), in which case the thermal temperature of the electrons often suffices to characterize the source function, which then is equal to the Planck function B_ν. There are situations for which that does not hold that we will refer to as non-LTE (NLTE). They can happen when radiative excitation rates are different from collisional excitation rates and/or the radiation field is not B_ν. This can happen for some transitions in some atoms and not for others in the same location. NLTE effects can arise indirectly, since we will see in Section A.1 that source functions can be written in terms of level populations, but these depend on other level populations and bound-free processes that might be influenced by radiation fields at rather different wavelengths from the transition under consideration. A radiative imbalance in one transition can affect the lower or upper level for a different transition. If LTE is violated then inferences based on observations can be seriously in error depending whether one is using the correct physics or assuming LTE.

Returning to the concept of optical depth, the "surface" of the star can be defined as the location where the vertical optical depth (from outside looking downward) is about unity. It is dependent on frequency, of course, and in some cases a star can have a significantly different "size" in some spectral lines than in others or the continuum, or in different frequencies of the continuum. To avoid this confusion one can define a frequency-integrated quantity called the Rosseland mean optical depth (Rutten section 4.2.2; we don't need the details here) and the bottom of the photosphere is often defined at the depth where the Rosseland optical depth has the value of two-thirds. It suffices to think of this as the place where the typical photon has about a 50% probability of flying free from the star. Anywhere above this depth that has a significant opacity is part of the stellar atmosphere.

From the observer's point of view, the slant optical depth indicates how far into the star one can see (at a given frequency and angle). The likelihood of seeing into a layer with vertical optical depth τ_ν is about $\exp(-\tau_\nu/\mu)$. For completeness it is also worth noting that if one is looking through an optically thin layer with no emissivity to a background intensity of I_B then the background intensity will be reduced by that same factor, which is typically assumed for ISM cases. Otherwise there will some replenishment of the diminished value of I_B based on the emissivity of the layer. On the other hand, if there is emissivity but no background intensity and the optical depth is significantly less than unity, then the observed intensity will be about $\tau_\nu(\mu)S_\nu$ (assuming a constant source function).

A parcel inside the atmosphere will receive radiation from all directions, but the radiation that reaches it will be weighted most strongly from regions that are near optical depth unity away in each direction. There won't be much contribution from nearby regions that are nearly transparent (low optical depth) or hidden far away (high optical depth) for the reasons in the previous paragraph. One can therefore define a "contribution function" which is essentially the kernel of Equation (A.3) since the same reasoning applies at the surface. The information contained in an observation at a given frequency is most relevant to the region of the atmosphere where the contribution function is high.

Above the bottom of the photosphere the bulk of the energy will escape by radiation even if convection dominates below (it becomes easy to transport energy by radiation when the remaining opacity is low). The condition of radiative equilibrium (Rutten section 7.3.2) employs the fact that the less opacity the next layer outward has, the shallower the temperature gradient has to be in order to move the same amount of total energy through that opacity. In general the overall opacity will decline outward because the atmosphere is getting less dense due to hydrostatic equilibrium, so the temperature gradient will be flattening. There is no reason for the temperature gradient to reverse and begin to increase outward in the radiative equilibrium case (that would be unphysical).

There are two basic scenarios to use in calculating this atmospheric structure. Rutten (chapter 7) provides details on how this is done in the simpler case of the "plane-parallel" atmosphere. This case is appropriate if the stellar atmosphere has a small vertical extent compared with the stellar radius. It models the star as a set of constant infinite planes in the horizontal directions and we calculate the evolving vertical structure of one plane to the next. In this theoretical case the inward half of the sphere around any point continues to be filled by radiation from the photosphere no matter how far up we go (because the planes are infinite). This means the radiative equilibrium temperature solution will flatten to some asymptotic value at the top of the atmosphere. The model is no longer physically appropriate if we move far enough away from a star that it doesn't fill essentially half the sky.

In the more realistic but complex spherical case, as we move away enough from the star radiation from the photosphere begins to fill less and less of the sky (the star begins to have an angular size less than π radians). The flux that has to be moved through a unit area now falls with the inverse square of the distance to the star so the radiative equilibrium temperature falls toward zero if we calculate out far enough. The plane-parallel case is reasonable to use for the atmospheres of main sequence stars. Its failure lies in the assumption that its physical properties are constant in all horizontal directions. That isn't true on real stars because of convection and magnetic fields.

Despite the firm conclusion that radiative equilibrium models must always have a temperature gradient that decreases with height (or with decreasing optical depth) above the bottom of the photosphere, real stars like the Sun instead often exhibit a reversal of the temperature gradient from decreasing outward to increasing outward at some higher point. That implies that there is a temperature minimum. We refer to the atmosphere inward of that minimum to the bottom of the photosphere as "the photosphere" (Chapter 2). The atmosphere above that point, where the temperature increases outward for some vertical extent, is called "the chromosphere" (Chapter 3). The Introduction explained that name arises because this part of the solar atmosphere appears red during a total eclipse, due to the fact that a major part of its visible light comes from $H\alpha$ emission. The only way such an atmospheric structure can occur is that there must be some non-radiative heating (radiative equilibrium is violated). The top of the chromosphere is defined where the temperature gradient takes a sudden jump to a much steeper gradient (increasing very quickly outward). That part of the atmosphere is discussed in Chapter 4.

A.1 LTE Line Formation

Because the chromosphere is above the photosphere, the continuum optical depth in the chromosphere is generally quite small. There are exceptions in a few optically thick continua, especially the Lyman continuum shortward of 91.2 nm, and less obviously in the mm continuum due to H^- and electron free–free opacities. These examples set the stage for a discussion of the complications in understanding chromospheric diagnostics: the Lyman continuum is subject to NLTE effects that must be carefully calculated, while the mm continuum is formed in LTE and thus its source function is just the Planck function at around optical depth unity (at the observing frequency). A succinct discussion of chromospheric observations in continua at long wavelengths can be found in section 4.11 of Linsky (2017). Most of the observational work on the chromosphere is done using spectral lines, which are much optically thicker than most of the continuum, and most of those are formed in NLTE.

To understand why NLTE effects arise and why they are important, let us work through the two-level atom case. For now we will ignore its bound-free continua so the only interaction of the atom with its surroundings is through transitions up or down between the lower level and the upper level. These transitions can take place either through absorption or emission of photons that have an energy close enough to the energy difference between the levels, or through collisions (usually with electrons) that add or subtract that amount of energy. Our first concern is with the probability of each of these energy exchanges, and we will think of those in terms of the rate at which each occurs. In a steady state the populations of the two levels will be determined by statistical equilibrium of the rate equations for the two levels. We will slowly add sophistication to this picture.

The intrinsic rates at which transitions occur for a given pair of levels in a given atom can be written using the Einstein coefficients (Rutten section 2.3). If the atom is in the upper level, it might spontaneously decay to the lower level through the emission of a photon; the rate of spontaneous de-excitation is denoted A_{ul}. Einstein realized that in order for everything to balance in thermodynamic equilibrium, there also has to be a way for the atom to be "tickled" into de-exciting by the passage of a photon with the requisite energy. This is called "stimulated emission" and it has the rate B_{ul}. It is this process that is responsible for laser and maser emission; the emitted photon has many characteristics of the stimulating photon (including its frequency and direction). Finally, the atom can be excited from the lower to upper level by the absorption of a photon of the appropriate energy with the rate coefficient B_{lu}. There are similarly rates for collisional excitation and de-excitation: C_{lu} and C_{ul}. All the actual rates except that involving A_{ul} depend on the radiation field or the electron density and temperature.

We now introduce some additional sophistication needed to understand the formation of spectral diagnostics. Because of quantum effects, the energy of the atom in the upper state will have some uncertainty. This is set by the value of A_{ul}; the longer it takes the atom to de-excite the sharper (more constrained) the possible energy states of the level (a form of the uncertainty principle). For our two-level atom the lower state will have a sharp (ground state) energy, since it never spontaneously de-excites. Real transitions that arise from atomic ground states are called "resonance lines." Many of the important chromospheric diagnostics are of this type because they are the most optically thick since most atoms will be in the ground state. The "fuzziness" of the upper state leads to a probability profile for absorption: ϕ_ν, and an emission profile that could be different: ψ_ν. These are often called "line" profiles because spectrometers operate using slits that look like lines.

In essence the atom behaves like a damped oscillator with damping constant of $\gamma^{rad} = A_{ul}$. The probability that the atom will decay away from the central energy difference (at ν_0) between the levels is the Lorentz profile: $\phi(\nu - \nu_0) = (\gamma^{rad}/4\pi^2)/((\nu - \nu_0)^2 + (\gamma^{rad}/4\pi)^2)$. This will also be the form of the emission profile ψ_ν for spontaneous de-exitation, but won't necessarily hold for stimulated emission or decay from an upper level that has been excited under special circumstances (this matters for some resonance lines of interest). Collisions can also disturb the atom while excited and they act like another form (actually several other forms, Rutten section 3.3) of damping. The Lorentz profile holds when the damping constants are simply added together, so one can calculate it with the sum of all forms of damping: $\gamma^{total} = \gamma^{radiative} + \gamma^{elastic\ collisions} + \gamma^{Stark} + \gamma^{Van\ der\ Waals}$.

In addition to this form of intrinsic broadening, the effective line profile for a spectral line is also affected by Doppler shifting. If another atom emits a photon at the central frequency but is moving with respect to the atom that will absorb it, the photon will be absorbed at an altered frequency (with a probability given by the intrinsic line profile). Motions that come into play include the thermal motion of the atoms and turbulence on scales comparable to the optical depth. For motions with a Gaussian velocity distribution (all but bulk motions), the broadening profile will look like $\phi_{Gaussian}(\nu - \nu_0) = (1/\sqrt{\pi}\,\Delta\nu_D)\exp((\nu - \nu_0)/\Delta\nu_D^2)$, where $\Delta\nu_D = \xi\nu/c$ and ξ is the velocity at the peak of the Gaussian.

The combination of these two broadening functions leads to the final line profile, given by the Voigt function (Rutten section 3.3.3). This looks like a Gaussian in the line core and has damping wings that decrease as ν^{-2}. Because the Gaussian core is really composed of shifted versions of line center, it will dominate out to frequencies of nearly $3\Delta\nu_D$. This brings us to low enough optical depth for typical photospheric lines that the continuum begins to take over, so the observed line shapes are basically Gaussian. For lines that are optically thick enough to be important for the chromosphere, however, the damping wings can be quite important. Doppler shifts due to turbulence on large scales, convective motions, and possible downdrafts, jets, or winds in the atmosphere along with the stellar rotation should be applied for the observer after the computation of the local line profile. The collisional broadening can dominate over the Doppler broadening when the densities are high enough (as is true for some spectral lines in red, brown, and white dwarfs). For carefully selected lines the effect on their wing widths of the stellar surface gravity provides an effective way to distinguish between main sequence stars, subgiants, and red giants.

The emission and absorption line profiles are usually equal to each other for either or both of two different physical reasons. One is if the lower state is also "fuzzy" (not a ground state) in which case the atom can excite or de-excite with slightly different energies (starting and ending at different places in the line profile). The second is when the atom is "jostled" by collisions while excited, which can slightly alter the de-excitation energy. The situation where the profiles are equal is called "complete redistribution"; it is generally the correct assumption for spectral lines other than strong resonance lines. Its physical meaning is that absorption can occur at various frequencies according to the line profile, and re-emission can likewise occur at various frequencies with the same probability. A spontaneously emitted photon will not remember at what frequency a photon was absorbed to excite the upper level. This subject is dealt with in more detail in Rutten section 3.4.

The assumption of complete redistribution means we can use ϕ_ν as the line profile for both emission and absorption processes, so it divides out of the rate equations. It is then straightforward to show that the source function at a given location is

$$S_\nu = j_\nu/\alpha_\nu = \left(n_\mathrm{u}A_\mathrm{ul}\right)/\left(n_l B_\mathrm{lu} - n_u B_\mathrm{ul}\right) \tag{A.4}$$

where the populations of the lower and upper levels are n_1 and n_u. Using the Einstein relations (which relate the Einstein coefficients to each other) this expression reduces to $S_\nu = B_\nu$ when the levels are populated as expected from the Boltzmann distribution in thermodynamic equilibrium. The question of whether we have LTE or NLTE thus comes down to the question of the ratio of the level populations. Over the line profile ν will be essential constant in the context of the Planck function, so we can treat the source function as frequency independent for the transition when we have complete redistribution.

The process of forming the great majority of absorption lines in a stellar spectrum for stars cooler than about 7000 K (around spectral type F or later) has now been explained. These lines are formed in the photosphere, where the temperature gradient decreases outward. Each line is most opaque at line center so the view there penetrates least into the star and the source function at optical depth unity is lowest. As we move out in the line profile the opacity decreases and the frequency-dependent optical depth unity moves further into the star where the source function is higher. Thus the line intensity is darkest at line center and grows brighter on both sides away from it according to the Voigt profile. It doesn't get a lot brighter while in the Doppler core (composed of shifted line centers) but sharply increases back to continuum values outside that unless the damping wings are strong enough. If one uses a very narrow-band filter that only samples a portion of the line profile, one can obtain an image of the star at the height where the optical depth is about unity at that frequency (this is only useful to do for the Sun).

One of the most common errors novices make is to think that spectral lines are darker than the continuum because some of the continuum radiation has been absorbed. It is perhaps unfortunate that most spectral lines are called absorption lines only because they are darker compared to the continuum. The physical meaning of absorption in radiative transfer has to do with the nature of the source function—it exchanges energy with the local thermal pool. Because of LTE the light seen at each frequency is produced locally where the frequency-dependent optical depth is near unity. It has nothing to do with the continuum intensity except at the extreme line wings. The line is simply optically thicker than the continuum so the continuum is hidden, and the line source function is lower. It is perhaps fortunate that despite this the source functions for most photospheric lines are in fact controlled by absorption rather than scattering.

The line continues to weaken (brighten) further from line center until the line profile reduces the opacity to no greater than the continuum opacity. We have then reached the bottom of the photosphere at that frequency and see the continuum intensity. The observed intensity then stays constant further from line center since the continuum opacity is changing very slowly with frequency. Note that this continuum opacity will typically have nothing to do with the line-producing atom. We have mostly neglected a discussion of the physical sources of the continuum in stellar photospheres; a brief discussion of that topic can be found in Rutten section 8.3 and a thorough one in most books on stellar atmospheres. For the majority of stars we are talking about the visible and infrared continuum opacity arises from H^-, which is formed in LTE. UV continua arise primarily from bound-free transitions of hydrogen or various metals and their ions. These often are not formed in LTE because they depend on the level populations of the atoms.

The physical processes of LTE line formation mean in principle that one can infer the temperature and density structure of most stellar atmospheres straightforwardly from many of their line profiles, as well as other physical quantities of interest like surface gravity, turbulent velocities, or chemical abundances. That will not be true for the strongest lines because LTE stops holding. Hotter stars than spectral type F exhibit photospheric lines that arise predominantly from hydrogen and helium. Their formation typically does not satisfy LTE and that makes them harder to extract detailed physical information from. We next move to the formation of NLTE lines.

A.2 NLTE Line Formation

Up to now we have mostly been assuming that the level populations are determined by LTE (have a Boltzmann distribution). In essence that either ignores radiative rates (which is fine if collisions dominate) or assumes they are the rates driven by the local Planck function. If we are concerned about the actual radiation field that drives the local radiative rates we have to include the rates due to radiation from all directions and over the frequencies relevant to the spectral line. To account for all directions we must use the mean intensity J_ν, which is simply the integration of I_ν over the whole sphere at a given location. Further integrating the radiation field over the relevant frequencies, the equation of statistical equilibrium for transitions between the two levels can then be written:

$$n_l\left(C_{lu} + B_{lu}\left[\int J_\nu\phi_\nu d\nu\right]\right) = n_u\left(C_{ul} + A_{ul} + B_{ul}\left[\int J_\nu\phi_\nu d\nu\right]\right). \qquad (A.5)$$

If we now substitute this relation into the source function written in terms of level populations (Equation (A.4)) and apply the Einstein relations while ignoring stimulated emission (generally fine for the stellar cases we are considering), the source function can be re-written in the form:

$$S_\nu = (1 - \epsilon)\bar{J} + \epsilon B_\nu; \text{ where } \bar{J} = \left[\int J_\nu\phi_\nu d\nu\right] \text{ and } \epsilon = C_{ul}/(C_{ul} + A_{ul}). \qquad (A.6)$$

One can now view the source function as composed of two parts: the first term is the scattering source function and the second term is the thermal (absorption) component. This form of the source function makes it clear that we will have LTE at a given location when C_{ul} is much larger than A_{ul} and/or $\bar{J} = B_\nu$. The reason the first condition works is that collisions dominate the rates that determine the level populations and those are governed by the local temperature. The density has to be sufficient to create enough collisions to cause this dominance. On the other hand if ϵ is small but the radiation field at this frequency has the value of the blackbody at the local temperature, we will still get LTE. Remember that the radiation field is characteristic of the source function near optical depth unity, integrated over all directions, so that might indeed not match the local Planck function. Whether it does depends partly on how opaque the atmosphere is in the spectral line, and in part how rapidly the temperature varies as a function of line optical depth in all directions. It might not vary that much, since the radiative temperature gradient is set by continuum optical depths that change slowly over physical scales that can encompass many line optical depths.

Even if the radiation field does not match the local blackbody, whether that makes a difference is dependent on the local density through the factor ϵ. It is set by the ratio of collisional to radiative rates. The closer ϵ is to unity, the more collisions are in charge and the more likely LTE will hold regardless of the radiation field. On the other hand, if ϵ is very small (say 10^{-6}) then we are much more likely to be in an NLTE situation. This arises for strong resonance lines where A_{ul} might have values from 10^6 to 10^8 and optical depth unity at line center probes chromospheric densities that are much lower than photospheric densities. As a final exacerbating factor, since chromospheric temperatures are higher and resonance lines tend to be at higher frequencies relative to typical photospheric lines, B_ν is more sensitive to temperature differences so it is easier to have a radiation field that does not match the local B_ν.

When scattering is important (small ϵ) the source function is generally lower than it would be in LTE. This will of course influence the radiation field so that $\bar{J}$ will also be lower. These effects will be largest near the surface for two reasons. The first is because the density will be smaller near the surface. The second is due to the fact that scattering lowers the source function by transmitting the information that no radiation is shining down from above the surface to deeper layers. That is because of the angle integration inherent in $\bar{J}$ and the fact that scattering does not alter photon energies. The absorption part of the source function allows the local thermal pool to act as a source of new photons, while the scattering part relies on receiving radiation from elsewhere. This reasoning also makes clear that as one goes deeper into the atmosphere, the source function will eventually reach LTE (when the density becomes high enough and/or the information about the surface has been scattered away).

It is possible to derive analytic expressions for the behavior of the mean intensity (or source function) in simple cases (Rutten section 4.3). Without repeating the details here, in an isothermal atmosphere (constant B_ν over depth) the source function reduces to the form:

$$S_\nu = \left(1 - \left(1 - \sqrt{\epsilon}\right)\exp\left(-\tau_\nu\sqrt{3\epsilon}\right)\right)B_\nu. \tag{A.7}$$

The physical content of this equation is that at the surface the source function will be reduced to $\sqrt{\epsilon}\,B_\nu$, and that it will recover to B_ν at an optical depth much greater than $1/\sqrt{\epsilon}$. The optical depth at which LTE is recovered is called the "thermalization depth." This behavior can have serious effects for small ϵ. For example when $\epsilon = 10^{-6}$ the line center source function at the surface will be one-thousandth of B_ν and does not recover to B_ν until well over 1000 optical depths inward. The very deep spectral line produced in this case is extremely different from what is produced if LTE obtains throughout. In fact a spectral line observed from the assumed isothermal LTE atmosphere is invisible; the same source function is observed no matter what the optical depth is. A set of examples illustrating how this works is shown in Figure A.1.

As mentioned when we introduced the line profile, a more generalized consideration of atomic processes requires multiplying the absorption (lower–upper) term in the rate equation by ϕ_ν but the emission (upper–lower) terms by ψ_ν. For example, because resonance lines have sharp lower levels then if there is no collisional damping while excited the atom will remember the exact energy at which it was excited and re-emit a photon of that energy. It is clear in this case that the source function will vary with frequency through the line profile, and thus the observed spectral line will look different than it would if complete redistribution held (Rutten section 3.4). The case where $\phi_\nu \neq \psi_\nu$ is called "partial frequency redistribution" (the word "frequency" is often left out for both cases). We discuss in Section 3.2 under what physical conditions this occurs and what the consequences are. For now it is appropriate to comment that partial redistribution is important only for damping wings; Doppler broadening will scramble the intrinsic profile over the width of a Doppler core. Thus when the source function is different over the line wings the differences occur over a number of such Doppler widths. Of course it is precisely this sort of spectral line that is likely to have a low value of ϵ.

The highly simplified discussion in this section contains almost all the relevant underlying physical concepts that come into play when NLTE physics must be taken into account. In detail things get more complicated in a hurry when the atom has more than two levels and the population of a given level can depend on what happens in a whole network of levels connected to it through possible transitions at a variety of energies. Further complications arise because the rate equations also

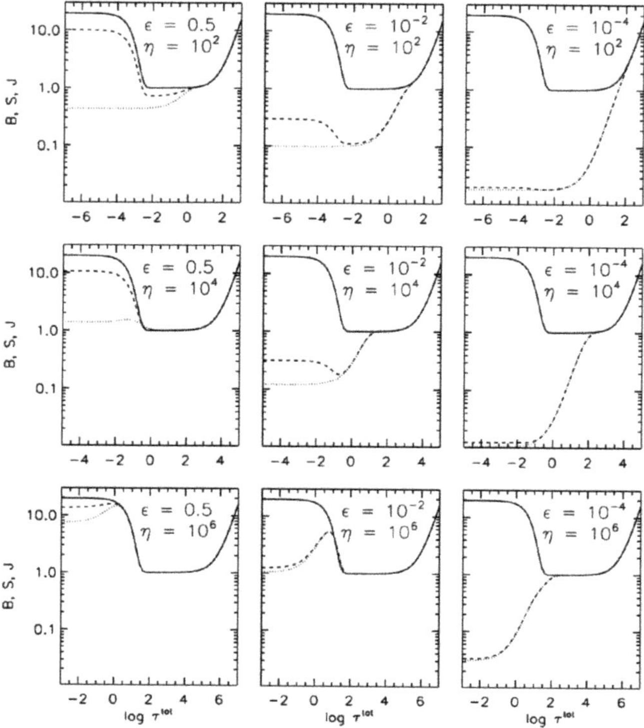

Figure A.1. A set of examples showing the behavior of the source function and mean intensity of a coherent scattering spectral line for various values of ϵ and various ratios η between the opacities of the line and its background continuum (which is in LTE). The continuum forces the source function to LTE at about $\log(\tau) = 2$. The chosen temperature structure that generates B has a chromosphere above a photosphere. It is clear in each row that the smaller ϵ gets, the deeper is the point above which the source function decouples from the Planck function. After decoupling the mean intensity contributes more to the source function as ϵ gets smaller and the line gets stronger compared to the continuum. This figure is borrowed directly from Rutten (figure 4.11). Credit line: Reprinted with permission R. Rutten. Copyright (c) 1995 Robert J. Rutten, Sterrekundig Instuut Utrecht, The Netherlands.

include the bound-free continua; those respond to radiation fields that are mostly at very different frequencies with very different optical depths than the line of interest. Furthermore the level populations are in play for both the source function and the optical depth scale. A particularly instructive example of how subtle effects can lead to unexpected consequences can be found in the explanation of photospheric emission lines from Mg I near 12 microns by Carlsson et al. (1992).

This makes it clear why LTE is an assumption that is abandoned very reluctantly. In LTE the source function is always the local Planck function and we are done if we know the temperature structure as a function of optical depth. Similarly, the optical depths are easier to compute if we know the density structure (which can be computed using hydrostatic equilibrium) since the level populations only depend on the local density and temperature. The problem then reduces to determining the temperature structure. That is quite tractable if one assumes radiative equilibrium and has a good knowledge of all the opacities involved. A number of grids of such models are available; one of the classic sets was started by Bob Kurucz (cf Buser & Kurucz 1992; Castelli & Kurucz 2004).

References

Buser, R., & Kurucz, R. L. 1992, A&A, 264, 557

Carlsson, M., Rutten, R. J., & Shchukina, N. G. 1992, A&A, 253, 567

Castelli, F., & Kurucz, R. L. 2004, A&A, 419, 725

Linsky, J. L. 2017, ARA&A, 55, 159

Rutten, R. J. 2003, Radiative Transfer in Stellar Atmospheres (Utrecht: Utrecht Univ.)

9 780750 321334